The Sunken Billions Revisited

ENVIRONMENT
AND
SUSTAINABLE DEVELOPMENT

The Environment and Sustainable Development series covers current and emerging issues that are central to reducing poverty through better management of natural resources, pollution control, and climate-resilient growth. The series draws on analysis and practical experience from across the World Bank, partner institutions, and countries. In support of the United Nations Sustainable Development Goals (SDGs), the series aims to promote understanding of sustainable development in a way that is accessible to a wide global audience. The series is sponsored by the Environment and Natural Resources Global Practice at the World Bank.

Titles in this series

The Changing Wealth of Nations: Measuring Sustainable Development in the New Millennium

Convenient Solutions to an Inconvenient Truth: Ecosystem-Based Approaches to Climate Change

Environmental Flows in Water Resources Policies, Plans, and Projects: Findings and Recommendations

Environmental Health and Child Survival: Epidemiology, Economics, and Experiences

International Trade and Climate Change: Economic, Legal, and Institutional Perspectives

Poverty and the Environment: Understanding Linkages at the Household Level

Strategic Environmental Assessment for Policies: An Instrument for Good Governance

Strategic Environmental Assessment in Policy and Sector Reform: Conceptual Model and Operational Guidance

The Sunken Billions Revisited: Progress and Challenges in Global Marine Fisheries

The Sunken Billions Revisited

Progress and Challenges in Global Marine Fisheries

 **WORLD BANK GROUP**

ISBN (paper): 978-1-4648-0919-4
ISBN (electronic): 978-1-4648-0947-7
DOI: 10.1596/978-1-4648-0919-4

Cover design and illustration: Bill Pragluski of Critical Stages

Library of Congress Cataloging-in-Publication Data has been requested.

Contents

Figures

Tables

Acknowledgments

This report was produced by a World Bank team led by Mimako Kobayashi and Charlotte de Fontaubert, and composed of Juan Jose Miranda and Carter Brandon. An initial draft of the report was prepared by Professor Ragnar Arnason (University of Iceland), who also developed the model underpinning the study.

Many individuals and institutions assisted in the development of this report. The Food and Agriculture Organization of the United Nations (FAO), University of Iceland, and University of California, Santa Barbara, provided important institutional support. Special thanks go to Birgir Thor Runolfsson at the University of Iceland and FAO who compiled the first draft of the chapter on trends in global fisheries, and to Rebecca Metzner and Anika Seggel at the FAO who provided invaluable assistance in obtaining fisheries data from FAO databases and other sources. Also gratefully acknowledged are the important contributions of the following individuals: Rashid Sumaila at the Fisheries Center of the University of British Columbia; Gordon Munro at the University of British Columbia; James Anderson at the University of Florida; Chris Costello and Tracey Mangin at the University of California, Santa Barbara; Matt Elliott and Emily Peterson at California Environmental Associates (CEA); Trond Bjorndal at the University of Bergen; James Wilen at the University of California, Davis; and Miguel Castellot at the DG MARE European Commission.

Guidance was provided by peer reviewers Arni Mathiesen (FAO), Flore Martinant de Preneuf (World Bank), Gordon Munro (University of British Columbia), Mitsutaku Makino (Japan Fisheries Research Agency), Nobuyuki Yagi (Tokyo University), Rebecca Metzner (FAO), and Trond Bjørndal (University of Bergen). Finally, the team is grateful to Paula Caballero and Valerie Hickey (World Bank) for their encouragement and technical insights for the entire duration of the project.

The Global Program on Fisheries (PROFISH) Multi-Donor Trust Fund supported the preparation and publication of the report.

About the Authors

Ragnar Arnason is a leading fisheries economist and a professor in the Faculty of Economics at the University of Iceland. He has been an active researcher in the field of fisheries economics for more than three decades and has published a large number of scientific articles and books in the field. He has been an adviser to many countries around the world on fisheries management issues. Dr. Arnason developed the bio-economic model used to estimate the economic potential of the global marine fishery, conducted the necessary empirical estimations and calculations, and prepared the initial draft report on which this final report is based.

Mimako Kobayashi is a senior natural resources economist at the World Bank. She has written extensively on bio-economic modelling applied to various natural resource management problems in the United States and around the world. At the World Bank, she contributes to the work of the Environment and Natural Resources Global Practice by applying the analytical rigor of applied microeconomics to a range of development issues. She co-authored the World Bank report, *Fish to 2030,* and supports fisheries operations projects, notably the West Africa Regional Fisheries Program (WARFP). She worked with Dr. Arnason to produce the initial draft report and led the World Bank team that produced the final report.

Charlotte de Fontaubert is a senior fisheries specialist at the World Bank. Her work supports operations for the development of sustainable fisheries world-wide, with a particular emphasis on East Asia and the Pacific, the Middle East, and North and East Africa. She is the author of numerous articles and global studies on marine biodiversity, international fisheries, and the natural resources of the high seas. She edited the final version of the report.

Abbreviations

CPI	Consumer Price Index (U.S.)
EEZ	Exclusive Economic Zone
EU	European Union
FAO	Food and Agriculture Organization (of the UN)
FBS	food balance sheet
FFA	foreign fishing agreement
FIPS	Fisheries Information and Statistics Branch (of the FAO)
IUU	illegal, unregulated, and unreported (fishing)
MEY	maximum economic yield
MSY	maximum sustainable yield
RFMO	Regional Fisheries Management Organization
UN	United Nations
UNCLOS	United Nations Convention on the Law of the Sea

All dollar amounts are in U.S. dollars, unless otherwise noted.

The Sunken Billions Revisited: Progress and Challenges in Global Marine Fisheries

Global marine fisheries are in crisis. The proportion of fisheries that are fully fished, overfished, depleted, or recovering from overfishing increased from just over 60 percent in the mid-1970s to about 75 percent in 2005 and to almost 90 percent in 2013 (figure O.1). Biological overfishing has led to economic overfishing, which creates economic losses.

An earlier study estimated annual lost revenues from mismanagement of global marine fisheries at $51 billion in 2004

To quantify the value of this economic loss, in 2009 the World Bank and the Food and Agriculture Organization of the UN (FAO) published a study on the economic performance of global fisheries, *The Sunken Billions: The Economic Justification for Fisheries Reform*. The study highlighted the very weak economic performance of the global fisheries sector, estimating the lost economic benefits at about $50 billion a year. This finding stimulated policy discussions and made a compelling case that comprehensive reforms were necessary in fisheries around the world to recover these sunken billions. The report also changed the direction of development assistance in support of international fisheries, including by the World Bank, which established reform of fisheries governance as the fundamental entry point to its fisheries investment programs.

The 2009 report was written in the context of a long-term decline in fish stocks, stagnant or even slightly declining catches since the early 1990s, and an increase

FIGURE 0.1
State of global marine fish stocks, 1974-2013

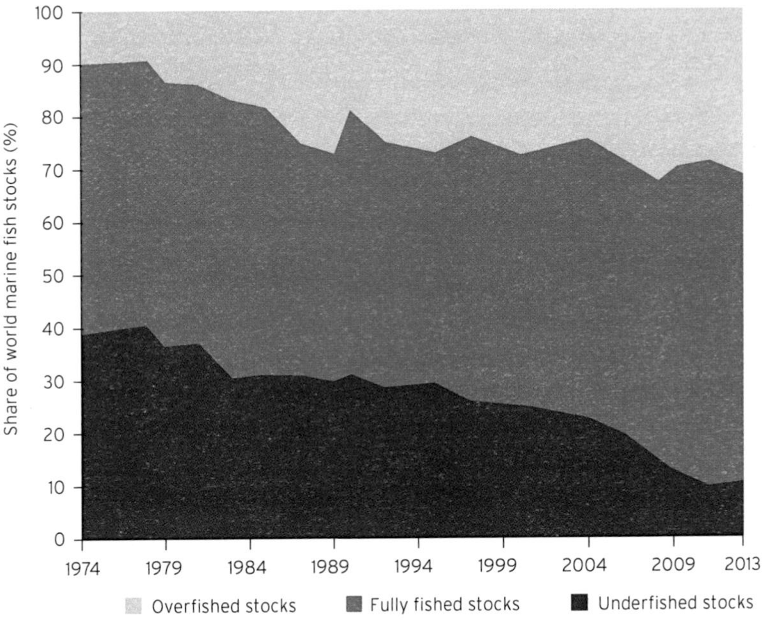

Source: FAO 2016.

in the level of fishing by a factor of as much as four. The productivity of global fisheries decreased tremendously, as evidenced by the fact that catches did not increase nearly as rapidly as the global level of effort (apparent in a doubling of the size of the global fleet and a tripling of the number of fishers). Another source of uncertainty is the increasing impacts of climate change, including sea-level rise, rising ocean temperatures, acidification, and changes in patterns of the currents.

This study follows the same approach as the earlier (2009) one. Both studies treat the world's marine fisheries as one large fishery, and they both model the economic performance of the sector in terms of this single aggregate fishery. This study, however, adds to the original one by deepening the regional analysis.

In addition, this study examines the range of complex issues that surround the reform of global fisheries management, including the financial and social costs of transitioning to a more sustainable resource management path, the considerable governance challenges associated with managing the largely open-access ocean resources, and the complicating factor of climate change. Although it does not attempt to address all of these issues fully, it lays out a comprehensive estimate of what the economic benefits of transitioning to higher value-added and more sustainable fisheries might look like.

This study estimates annual lost revenues at $83 billion in 2012

The primary objective of this study is to reinforce the messages of the 2009 publication and to catalyze calls for accelerating and scaling up the international effort aimed at addressing the global fisheries crisis. The analysis reveals economic losses of about $83 billion in 2012, compared with the optimal global maximum economic yield equilibrium.

These sunken billions represent the potential annual benefits that could accrue to the sector following both major reform of fisheries governance and a period of years during which fish stocks would be allowed to recover to a higher, more sustainable, and more productive level. These stocks cannot be recovered immediately, even if ideal sector governance were somehow imposed overnight. Rather, the process of recovery implies large transition costs and long-term sector restructuring.

Restoring fisheries would yield substantial returns

Severely overexploited fish stocks have to be rebuilt over time if the optimal equilibrium is to be reached and the sunken billions recovered. To allow biological processes to reverse the decline in fish stocks, fishing mortality needs to be reduced, which can only happen through an absolute reduction in the global fishing effort (as captured by the size and efficiency of the global fleet, usually measured in terms of the number of vessels, vessel tonnage, engine power, vessel length, gear, fishing methods, and technical efficiency). Reducing the fishing effort in the short term would represent an investment in increased fishing harvests in the longer term. Allowing natural biological processes to reverse the decline in fish stocks would likely lead to the following economic benefits:

- The biomass of fish in the ocean would increase by a factor of 2.7.

- Annual harvests would increase by 13 percent.

- Unit fish prices would rise by up to 24 percent, thanks to the recovery of higher-value species, the depletion of which is particularly severe.

- The annual net benefits accruing to the fisheries sector would increase by a factor of almost 30, from $3 billion to $86 billion.

This study looks at two hypothetical pathways that would allow fish stocks to recover. At one extreme, if the fishing effort were reduced to zero for the first several years and then held at an optimal level, global stocks could quickly recover to over 600 million tons in 5 years and then taper off toward an ideal

FIGURE 0.2
Incremental benefits of global fisheries reform: Projected dynamics of biomass

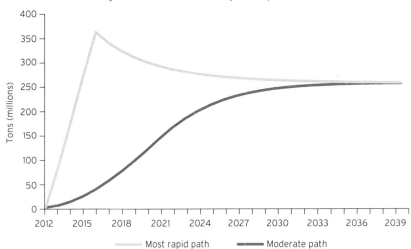

Most rapid path   Moderate path

Note: This graph shows incremental benefits above the estimated biomass baseline of 215 million tons in 2012. The most rapid path involves reducing fishing effort to zero for first several years and then holding it at the optimal level. The moderate path involves reducing global fishing effort by 5 percent a year between 2012 and 2022.

level. Reducing the global fishing effort by 5 percent a year for 10 years would allow global stocks to reach this ideal level in about 30 years (figure O.2).

The need for reform is greatest in Asia and Africa

This study extends the original investigation to identify the economic performance of fisheries in five world regions (Africa, the Americas, Asia, Europe, and Oceania). Because initial economic performance and the level of overexploitation vary greatly by region, the effort required would differ across regions (figure O.3).

The quality of data varies greatly across regions, rendering the assessments of economic performance and the estimates of forgone economic benefits by region less reliable than the global results. The regional results should be interpreted in that light and a continued effort made to improve fisheries statistics at the national, regional, and global levels.

Transitioning to a sustainable level of fishing would be difficult—but the benefits would far exceed the costs

Transitioning to a sustainable level of fishing would involve significant policy and governance challenges at the global, national, and local levels. It would also impose costs on some stakeholders. The single largest source of economic gain from moving to a sustainable level of fishing would be the reduction in fishing costs

FIGURE 0.3

Distribution of sunken billions, by region

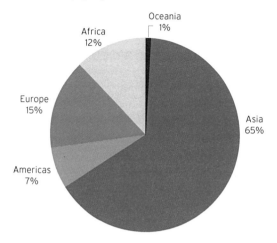

FIGURE 0.4

Sources of economic benefits from moving to the optimal sustainable state for global fisheries

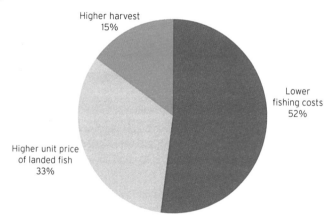

(figure O.4). This reduction, however, would impose very high adjustment costs on both the fishing industry and the upstream and downstream industries and services, with displaced vessel owners and fishers bearing the brunt of the costs.

Climate change will have additional negative impacts on global marine fisheries, calling for quicker action

Sea-level rise, higher ocean temperatures, increasing acidification, and changes in the ocean current patterns will all have tremendous impacts on global fish stocks and the related ecosystems, in ways that are not yet fully understood

(Alison and others 2009). They add a sense of urgency to long overdue fisheries reforms, because they threaten the ability of depleted stocks to recover from overexploitation, as they had done in the past.

Reform will require financial and technical assistance at many levels

This report makes a very clear case for the need for reform. It does not analyze policies, financing, or the socioeconomic impacts of embarking on such reform.

Many case studies have shown that different strategies are called for in different circumstances (Worm and others 2009). Whichever strategies are chosen, fishing capacity will have to be reduced, jeopardizing the livelihoods of millions of fishers. Financing will be needed to fund the development of alternatives for them, to provide technical assistance at all levels, and to conduct additional research on ecosystem changes and related ecological processes.

This report poses important questions. If the sunken billions wasted annually at sea are to be recovered, and fisheries put on a sustainable pathway, policy makers will need to answer these questions, and soon.

References

Alison, E., A. Perry, M. Badjeck, W. Adger, K. Brown, D. Conway, A. Halls, G. Pilling, J. Reynolds, N. Andrew, and N. Dulvy. 2009. "Vulnerability of National Economies to the Impacts of Climate Change on Fisheries." *Fish and Fisheries* 10 (2): 173–96.

FAO (Food and Agriculture Organization of the UN). 2014. *The State of World Fisheries and Aquaculture 2014 (SOFIA)*. Rome: FAO.

———. 2016. *The State of World Fisheries and Aquaculture 2016 (SOFIA)*. Rome: FAO.

World Bank and FAO (Food and Agriculture Organization). 2009. *The Sunken Billions: The Economic Justification for Fisheries Reform*. Washington, DC: World Bank.

Worm, B., R. Hilborn, J. Baum, T. Branch, J. Collie, C. Costello, M. Fogarty, E. Fulton, J. Hutchings, S. Jennings, O. Jensen, H. Lotze, P. Mace, T. McClanahan, C. Minto, S. Palumbi, A. Parma, D. Ricard, A. Rosenberg, R. Watson, and D. Zeller. 2009. "Rebuilding Global Fisheries." *Science* 235: 578–85.

CHAPTER 1

Introduction: Trends in Global Fisheries and Fisheries Governance

The original *Sunken Billions* (2009)

In 2009, the World Bank and the Food and Agriculture Organization of the UN (FAO) published a study on the economic performance of global marine fisheries, titled *The Sunken Billions: The Economic Justification for Fisheries Reform* (World Bank and FAO 2009). This global analysis estimated net benefits that the fishing sector globally was generating in 2004, compared to what it could have potentially generated sustainably. The study highlighted the dismal economic performance of global marine fisheries and made a strong case for paying due attention to the economic health, as well as biological health, of the world's fisheries. The study reported estimates of around $51 billion in economic losses in 2004 compared to the benefits of a more sustainable global fisheries management regime.[1]

The staggering magnitude of these estimated foregone economic benefits, or sunken billions, stimulated a policy discussion that comprehensive reforms were necessary in many fisheries around the world. It also changed the direction of development assistance, including that of the World Bank, for international fisheries, by putting fisheries governance reform (as opposed to fishing capacity expansion) as the fundamental entry point of its fisheries investment programs.

The original *Sunken Billions* report used data until 2004, a time of long-term decline in fish stocks and falling productivity. More than a decade later, and despite a few positive indicators (mainly associated with improved fisheries governance in several important fisheries in the Americas, Europe, and

Oceania), deterioration of biological and economic health persists in many individual fisheries. As a result, poverty in coastal fishing communities remains a major development agenda in many coastal and island developing countries. In addition, in the face of worsening and new impacts of climate change, there is growing uncertainty regarding the state of fish stocks and marine ecosystems more broadly. While sea-level rise directly threatens the way of life of coastal populations, rising ocean temperature, acidification, and changes in oceanic current patterns also affect how much fish are found, and where, in the oceans.

Approach and scope of this updated study

In this context, the present study updates the previous estimate of foregone economic benefits of global marine fisheries using 2012 data, which is the most recent year for which reasonably complete data are available. By reporting a new set of estimates of the economic performance of world fisheries, the primary objectives are to reinforce the messages of the 2009 publication and to catalyze calls for further action to accelerate and scale up the international effort to improve global fisheries.

This study follows the same modeling approach as the original study, which is to regard global marine fisheries as one large fishery and to conduct an economic performance assessment for this aggregate fishery. (An alternate approach would be to model individual fisheries, or at least a reasonable sample of them, and subsequently aggregate the outcomes.) Here, a single bio-economic model that characterizes both the biological and economic aspects of the global fishery is used to simulate the outcomes of interactions between human action (that is, fishing) and biological forces of the fish population.

Representing the complexity and the dynamics of global fisheries in a single model involves a substantial simplification of reality. Potentially valuable information and the idiosyncrasies of individual fisheries are ignored in this approach. However, the simple, and yet theoretically consistent representation allows the analysis to focus on several fundamental drivers (most notably aggregate fishing effort) and key outcomes of marine fisheries systems (such as the biological state of fish stocks and the economic performance of the fishery). Accordingly, the results are transparent and relatively easy to interpret.

Scope of the study

Given the adopted analytical approach, it is useful to define the scope of this study. First, the study presents a set of objective assessments of the aggregate fishery's economic state and potential. However, it does not provide prescriptive policy recommendations of how to improve individual fisheries. While the model simulates dynamic outcomes of various fishing effort scenarios, it does not prescribe *how* fishing effort should be adjusted.

Second, the study discusses variations in the biological and economic health of regional fisheries as it extends the investigation to include the economic performance of marine fisheries in the main regions of the world: Africa, the Americas, Asia, Europe, and Oceania.[2] However, the discussions are limited to an objective performance assessment at the aggregate level; a discussion about issues and policies specific to each region is beyond the study's scope.

Third, there is limited discussion of the impacts on marine fisheries of various relevant and important issues, such as climate change and expansion of aquaculture production. Characterization of the global or regional aggregate fisheries in the bio-economic model that this study employs is reduced to a small number of input parameters. As a result, while attempts could be made to relate some specific consequences, for example, of climate change to model input parameters, such relationships are not clearly established and obtaining scientifically meaningful results would be difficult. This study does not factor in changing (growing) consumer demand for seafood, as the bio-economic model only depicts the dynamics of production relationships. Interactions between wild-caught and farm-raised fish in international seafood market are studied elsewhere (for example, World Bank [2013]).

Organization of the report

The main text in this report provides detailed explanations of the methodology and data used, and presents and discusses the main findings. To keep the text accessible to most readers, many of the more technical aspects of the model and inputs are included in a series of appendixes.

The report is organized as follows:

- Following this introduction, chapter 1 presents a review of recent trends in global fisheries to provide relevant context and introduces some key concepts that are used in the bio-economic model.

- Chapter 2 presents the basic methodology used to assess aggregate economic performance of global marine fisheries. In particular, the core functions of the bio-economic model and the model inputs are explained. The appendixes provide further discussion.

- Chapter 3 presents the main quantitative results: the estimate of the foregone economic benefits in the world's marine fisheries in 2012 and the associated confidence intervals. It presents the new results obtained from using the same model with newly available data, and a short discussion on the evolution of the performance of global fisheries between 2004 and 2012. The chapter also provides sunken billions estimates for the five regions of the world.

- Chapter 4 analyzes the dynamics of the global fisheries by comparing recovery model outcomes generated under alternative scenarios of aggregate fishing effort starting with the observed state in 2012. In particular, three fishing-effort scenarios are compared, as follows: (1) the most rapid path, where an economically optimum level of fishing effort is maintained following an initial fishing closure; (2) the moderate path, where global fishing effort is reduced by 5 percent annually until it reaches the same level as in (1); and, (3) the current path, where the fishing effort is maintained at the 2012 level. Finally, some possible paths for policy reform implementation are compared and discussed.

Current trends in global fisheries

Stocks and catches

In biological terms, the crisis in marine fisheries has been well documented. Globally, the proportion of fully fished stocks and overfished, depleted, or recovering fish stocks has increased from just above 50 percent of all assessed fish stocks in the mid-1970s to about 75 percent in 2005 (FAO 2007a), and to almost 90 percent in 2013 (FAO 2014a), as illustrated in figure 1.1. In FAO statistics, fish stocks are defined as fully or overfished if their biomass is at or below the level that supports maximum sustainable yield (MSY). Maximum economic yield (MEY), which maximizes the sustainable net benefits flowing from the stocks, occurs at a stock size that is larger than that at MSY level. Therefore, the FAO assessment of the biological state of the fish stocks in figure 1.1 indicates that

FIGURE 1.1
Global trends in biological states of fish stocks, 1974-2013

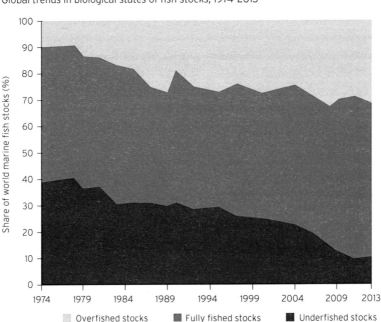

Source: FAO 2016.

FIGURE 1.2
Global marine catches 1950-2012

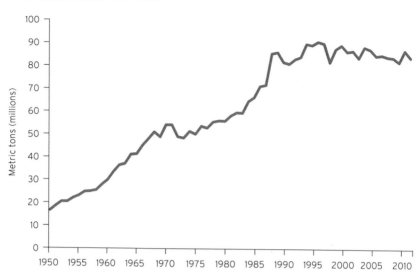

Source: FAO FishStat Plus database.

approximately 90 percent of the world's fisheries likely were subject to economic overfishing in 2011. Note, however, that these figures are based on the *number* of assessed fisheries, rather than the *volume* (the size of fish stocks or catches in the assessed fisheries), and thus do not necessarily measure the degree of biological or economic overfishing in relation to the global fish population.[3]

Figure 1.2 illustrates the evolution of global marine catches from 1950 to 2012. It shows an increasing trend in the reported global marine catch, lasting for about four decades from 1950 to the early 1990s. During this period, marine catches increased by approximately 1.4 million tons each year.[4] Since then, the reported global marine catch has largely stagnated, fluctuating between 79 and 86 million tons per annum. From the peak of 86 million tons in 1996, global marine catches have shown a small downward trend of about 0.2 million tons per annum. The data confirm the shift's statistical significance from a regime of growing catch to a regime of stagnation in the early 1990s.

This shift in the trend of global marine catches may be attributed to two major factors. The first possible factor is the general biological overexploitation of global fish stocks beyond the biomass level corresponding to MSY, which inevitably leads to subsequent reduction in catches. The second possibility is that some major fishing areas in the world may have reduced fishing efforts to allow depressed fish stocks to recover. Most likely, both factors have combined to put an end to the growing trend of marine catches.

Underlying trends in the species composition of the catch are behind the stagnation and slight decline in catches since the 1990s. Figure 1.3 shows a

FIGURE 1.3
Evolution of global marine catches by species group, 1950–2012

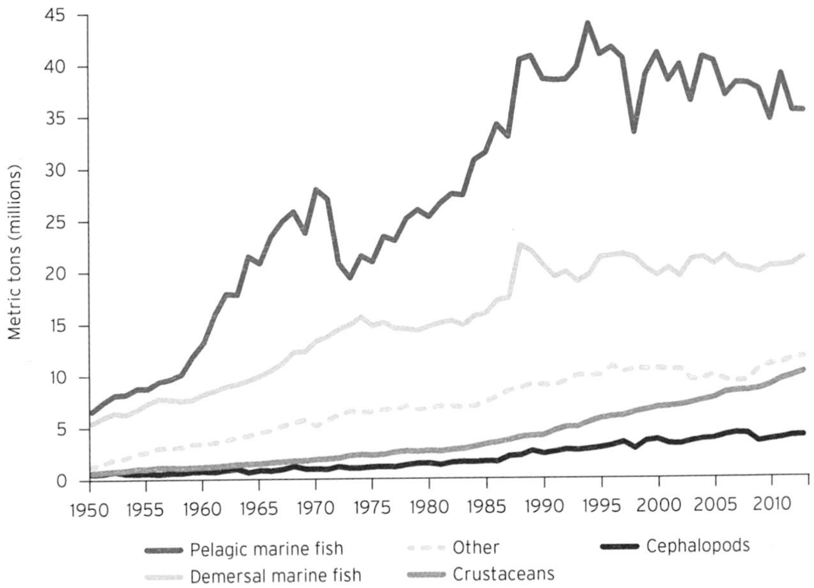

Sources: FAO 2014b; FAO FishStat Plus database.

considerable growth in the recorded catch of the demersal species (near-bottom dwelling) and the pelagic species (inhabiting the upper layers of the sea) between 1950 and 1970. After 1985, catches of demersal fish, the most valuable fish category, have stabilized at around 20 million tons per annum, while the catch of other species continued to evolve. The catches of pelagic species, which form the largest volume caught, grew to a peak of almost 44 million tons in 1994, but have since fluctuated between 35 and 41 million tons, with a small declining trend. The global catch of remaining categories has either showed a recent leveling-off (cephalopods and other species) or continuing increase (crustaceans) (FAO 2014b; FAO FishStat Plus database).

Fishing effort and productivity

While global marine catches stagnated and even declined, fishing effort appears to have increased. (Fishing effort is a composite indicator of fishing activity, including the number, type, and power of fishing vessels; the type and amount of fishing gear; the contribution of navigation and fish-finding equipment; and the skill of the skipper and fishing crew.) While the available global data on fishery inputs, both quantitative and qualitative, are limited and not always reliable, they all point in the same direction of greatly expanded fishing effort over the past 70 years (broadly speaking since the end of World War II).

First, according to FAO statistics, the reported global fleet has more than doubled over the past four decades, reaching a total of more than 4.7 million decked and undecked units in 2012 (FAO 1999; FAO 2014b), with Asia accounting by far for the highest number of decked and undecked vessels.

Second, the number of fishers (defined as the number of individuals worldwide engaged in catching fish, in either artisanal or more commercial-scale operations) in the sector appears to have grown even faster than the number of fishing vessels, which according to FAO (1999, 2014a) more than tripled over the past four decades. The average growth rate of the number of fishers has been almost 2.8 percent per annum, which is considerably higher than the growth rate of the world population, which peaked at 2.2 percent in 1963, and has declined since. The increase in the number of fishers is unevenly distributed around the world. From the 1970s, this increase mostly took place in low- and middle-income countries, while, in contrast, their number has been declining especially in most industrial economies, particularly over the past two decades. This decline can be attributed to several factors, including the relatively low remuneration in fishing—a sector often characterized by high-risk and difficult working conditions—growing investment in labor-saving technology aboard fishing vessels, and declines in fish stocks coupled with increasingly restrictive fisheries management measures (FAO 2007b).

Third, alongside this increase in vessels and labor employed in the global fishery, substantial advances in fishing technology occurred over the past four decades. This improved technology applies to vessels and various fishing equipment, including fishing gear and fish-finding devices. During the past four decades, technological improvements likely at least doubled and probably quadrupled the efficiency of fishing capital and labor that contributed to the fishing effort.

Thus, it is clear that there has been a substantial increase in the global fishing effort over the past four decades. Even conservatively estimated, this increase can hardly be less than fourfold.[5] Over the same period, however, the level of global marine catches has not even doubled, signifying that catch per unit effort, often considered as a measure of fishing productivity, must have fallen greatly. This situation is supported by the data: as illustrated in figure 1.4, the average reported harvest per capture fisher (marine and inland) declined by more than 50 percent, from just under 5 tons annually in 1970, to only 2.3 tons in 2012 (FAO 1999; FAO 2014a). It is noteworthy, however, that since 2000, the rate of decline in catch per fisher has greatly slowed and the decline has practically halted in the most recent years.

This decline in average output per fisher should be viewed in the context of the technological advances that have taken place in the world's capture fisheries over the same period. The pertinent technology includes large-scale motorization of traditional small-scale fishing boats, the increase in the use of active fishing

FIGURE 1.4
Average catch per fisher per year

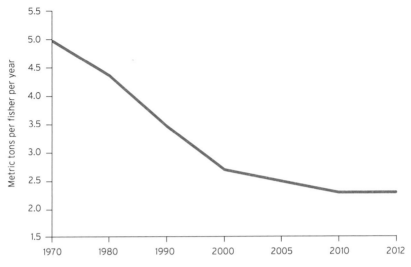

Sources: FAO 1999; FAO 2014a.

gear, such as trawling and purse seining, the introduction of increasingly sophis-
ticated fish-finding and navigation equipment, and the growing use of modern
means of communication. Although this technological progress has certainly
increased labor productivity in many fisheries, the overall negative trend seems
overwhelmingly driven by the increasing number of entrants into the sector (due
to poor governance), combined with decreasing catches (due to the depressed
state of fishery resources).

Fish prices

Figure 1.5 illustrates the evolution of the ex-vessel price of wild-caught fish and
the farmgate price of farmed fish since 2000 (FAO 2008; FAO 2014b). Also in
figure 1.5 is the price of all fish (wild caught plus farmed) in nominal and real
terms (in 2000 U.S. dollars).[6] The farmed fish price is consistently higher than that
of wild-caught fish due to the emphasis on the production of relatively high-value
species in aquaculture (for example, shrimp and salmon). In nominal terms, the
average price of wild-caught fish rose consistently, while the price of farmed fish
declined slightly during the first half of the 2000s, followed by a rapid increase.
The average real price of all fish was more or less constant from 2000 to 2006,
and has slightly increased since then.

    A number of factors are at play that can explain the observed trend in average
fish prices. First, the average price of wild-caught fish is linked to the state of
wild-fish stocks because this price determines at least partly the species composition of the catches. As seen previously in figures 1.2 and 1.3, there has been a

FIGURE 1.5

Estimated global average fish prices

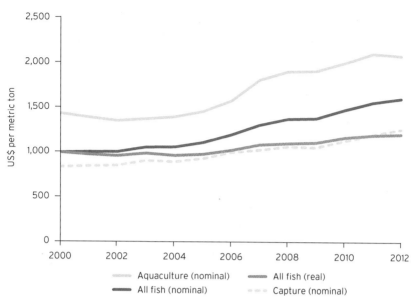

Sources: FAO 2008; FAO 2014b.

substantial expansion in global marine fisheries since the 1950s. The expanded production varies across different species groups. When fish stocks are commercially exploited, the most valuable stocks and larger individuals are typically targeted first. With this pattern applied over decades, global marine catches over time have comprised an increasing proportion of relatively less-valuable species and smaller fish. This situation has contributed to lowering the average price of landed catch compared to what would otherwise have been the case.

Second, another factor that contributes to depressing the average price of wild catches is the reduction of discards at sea and the increased landing of bycatch, which tend to comprise species of lower value. According to Kelleher (2005), globally the amount of catch discarded at sea decreased by over 10 million tons between 1994 and 2004. This reduction in discards can be explained both by technical improvements that have reduced unwanted catch and the greater proportion of the total catch that is now landed and used.

The wild-caught fish price is also influenced by the production of farmed fish, which has increased tremendously since the early 1980s, as illustrated in figure 1.6, which compares it to marine and inland capture fisheries production (FAO 2014b; FAO FishStat Plus database). As shown in figure 1.6, total 2012 aquaculture production was about 67 million tons, against 80 million tons for marine catches (FAO 2014a). However, a considerable part of the marine catches is low-quality fish, mainly used to produce feed (fishmeal and fish oil) for animal

FIGURE 1.6
Global fish production, 1950-2012

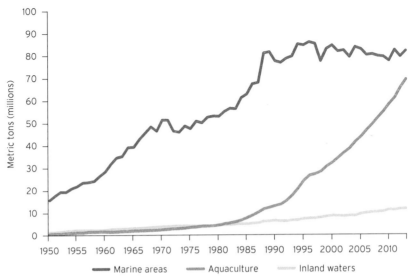

Sources: FAO 2014a and b; FAO FishStat Plus database.
Note: Aquatic plants are excluded.

production, including fish farming. In contrast, aquaculture accounted for about half of food fish supply for direct human consumption in 2012.[7]

This massive increase in the farmed fish supply has no doubt had a dampening impact on the price of wild-caught fish for direct human consumption. Since the increase in the farmed fish supply is greater than that of capture fisheries production, the impact on the wild-caught fish price was likely quite substantial.[8] On a limited level, the expansion of fish farming has also probably increased the price of wild fish for reduction purposes (that is, to produce fishmeal and fish oil, and mainly small pelagics). However, since the value of these landings is very small in relation to landings for direct human consumption, the impact of this effect on average wild-fish price likely remained relatively small.

Concurrently, several very different factors have contributed to influencing the upward climb of fish prices. In particular, the global consumption demand for fish products has been on the rise, driven chiefly by (i) population growth, (ii) higher incomes, especially in middle-income countries, and (iii) increased globalization of seafood markets. This increase in demand is illustrated by World Bank (2013) projections, which show substantial growth in fish consumption by 2030 in Africa, China, and India. Spurred by the globalization of markets for seafood, fish has become one of the most internationally traded agricultural commodities. For instance, in 2013–14, 36 percent of global fish production was traded in international markets (FAO 2015), and in 2012, fish trade accounted for 9 percent of global agricultural commodity trade (FAO FishStat Plus database).

Fisheries governance and management

Common ground in fisheries governance

As noted above, the overall performance of fisheries depends significantly on the state of the targeted stock or stocks, which, in turn, is directly affected by the fishing effort—both *how much* of it is exerted and *what type of fishing* takes place. How this effort is organized, and to a very large extent controlled and limited, falls under the purview of fisheries governance and management. Judging by the state of global fisheries, as illustrated here, and the extent to which overfishing has increased—even, in some documented instances, leading to fisheries collapses— the state of governance worldwide varies greatly and, despite some encouraging successful approaches, is in dire need of improvement.

One of the greatest and most vexing problems that plagues fisheries management is the open access regime, under which a common pool of fish resources can be accessed and harvested by anyone. This overwhelmingly leads to overcapacity and overexploitation (Gordon 1954; Hardin 1968; Ostrom 1990). In fact, economic theory predicts that in mature fisheries that are operated under such open access regimes, equilibrium profits tend to remain very small, at a level just sufficient to keep the fishers in the industry (Clark 1990; Gordon 1954), but generating little or no economic benefits.

While this open-access policy is widely recognized as a main driver of overfishing, however, many, including the FAO, also acknowledge that there is still an ongoing debate about the most effective and equitable way of authorizing access and allocating resources. [9] In fact, local, national, and regional circumstances vary widely, and while some approaches have proved successful in certain contexts (Arnason 2007; Arnason 2012; Costello, Gaines, and Lynham 2008; Costello and others 2010), developing analogous approaches that would be effective and acceptable in many of the different circumstances around the world remains a major challenge. Nevertheless, if governments want their fisheries sector to contribute sustainably to their national economies, they will need to invest in rebuilding biological overexploited resources and give fishers an opportunity to not only fish sustainably, but profitably as well. To that end, the issue of excess fishing capacity must be addressed, and fisheries must be effectively governed and managed.

According to the FAO, modern fishery governance is "a systematic concept relating to the exercise of economic, political and administrative authority," which is "characterized by:

- Guiding principles and goals, both conceptual and operational;
- The ways and means or organization and coordination;

- The infrastructure of socio-political, economic and legal institutions and instruments;
- The nature and modus operandi of the processes;
- The actors and their roles;
- The policies, plans and measures that are produced; as well as
- The outcomes of the exercise."[10]

By their very nature, fisheries governance regimes are complex and often need to be adjusted to respond to a variety of changes, including biological shifts that can be quite pronounced, or that merely try to address existing inefficiencies. In response to the evolving global fisheries crisis, governance has also evolved, and new, innovative approaches have been adopted, sometimes with great success. Indeed, studies have shown that, based on relative success measured in a variety of circumstances, combined fisheries and conservation objectives can be achieved by merging diverse management actions, including catch restrictions, gear modification, and closed areas, depending on the local context. In fact, "the feasibility and value of different management tools depends heavily on local characteristics of the fisheries, ecosystem, and governance system" (Worm and others 2009).

Fisheries governance regimes are very expensive to set up and operate, and their cost can vary depending on the type of conservation and management measures implemented. These costs range from scientific advice and management to enforcement—monitoring, control, and surveillance—and can reach 1 to 14 percent of the value of landings (Schrank, Arnason, and Hannesson 2003; Kelleher 2002). An additional issue is that, as highlighted in the original *Sunken Billions* report (World Bank and FAO 2009), only part of these costs are borne by fishers themselves, and a larger share falls on the public sector, while the benefits tend to be concentrated on the fishers who are proportionally much fewer.

In the face of these considerable and growing costs, many developing countries often find that it is difficult to finance properly this otherwise essential function; the few studies that were conducted on the issue suggest that there have been inadequately low levels of management expenditures (Willmann, Boonchuwong, and Piumsombun 2003). This is also true in the case of foreign fishing access agreements, where developing coastal states grant access to the resources in their waters to distant water fishing nation fleets, but do not collect sufficient payment in return for this access even to cover the costs that an effective fisheries management would incur (World Bank 2014).

The threat of illegal, unregulated, and unreported fishing

As noted above, the international legal fisheries regime has evolved into a series of treaties and legal instruments that clearly define the rights and obligations of states and fishing fleets (box 1.1), but this complex regime can only be effective if

> **BOX 1.1**
> The evolving international legal fisheries regime
>
> The international fisheries governance regime has evolved over centuries, going as far back as 1493, when there was an unsuccessful papal attempt made to divide and share the Atlantic Ocean between Spain and Portugal. Over hundreds of years, and ultimately due to global overfishing that resulted from technological improvements and increases in demand and the level of effort, a new regime was codified in the 1982 United Nations Convention on the Law of the Sea (UNCLOS). Under UNCLOS, coastal states exercise exclusive jurisdictional rights over 90 percent of marine living resources in so-called Exclusive Economic Zones (EEZs); the EEZs usually extend up to 200 nautical miles from their coastline. However, this relatively new jurisdictional reality does not conform to the biological reality that many fish species are migratory and some travel very long distances, sometimes between different EEZs and/or EEZs and the high seas. As a result, and after years of negotiations, UNCLOS was modified by the 1995 UN Agreement on Fish Stocks and Highly Migratory Fish Stocks, which calls on all states, both coastal and distant water fishing nations, to cooperate to manage these stocks sustainably through so-called Regional Fisheries Management Organizations (RFMOs). Underpinning the UN Fish Stocks Agreement and a series of additional international legal instruments are a very strong call for sustainable management of fisheries, the need to take an ecosystem approach to fisheries management, and the need to implement a precautionary approach.
>
> *Source:* de Fontaubert and Lutchman 2003.

the relevant actors agree to be bound by it, agree to comply with its requirements, and are willing to enforce conservation and management measures against any party unwilling to do so. By definition, these measures mark restrictions on use of marine living resources, often hindering the freedom of fishing that might have prevailed before. As long as all involved agree to abide by these measures, cooperation is facilitated, and success is more likely.

The regime's effectiveness is undermined, however, when a vessel, fleet, a state, or group of states unilaterally decides to ignore these measures or refuses to be bound by them. In the case of such a "free rider," the first beneficiary of the management is likely to be the delinquent actor who avoids any restrictive measures, all while the stock is otherwise protected or managed (more or less sustainably, as the case may be). More broadly, the compliance costs are borne by all the other parties that curtail their freedom to fish to comply with the measures adopted.

Therein lies the heart of the enforcement dilemma: the best conservation and management measures can be completely undermined at both the national and international level if free riders decide to take advantage of the compliance of others by engaging in illegal, unregulated, and unreported (IUU) fishing, as defined by the FAO International Plan of Action on the issue as follows:

- Illegal fishing refers to inter alia activities by a foreign vessel in the waters of a coastal state without its permission, and fishing activities by vessels flying the flag of a nonmember in waters regulated by an RFMO;

- Unreported fishing refers to fishing activities that have not been reported or that have been misreported (to the relevant national authority or in contravention of the procedures of the relevant RFMO); and

- Unregulated fishing includes activities in the waters regulated by an RFMO by a vessel without nationality or flying the flag of a state not party to that RFMO, and fishing in areas in which there are no applicable conservation and management measures, in violation of state responsibilities under international law.

This issue is of considerable importance and becoming more so. According to the first worldwide analysis carried out in a 2009 study, catches from illegal and unreported fishing alone accounted for as much as $23.5 billion annually, representing an estimated 11 to 26 million tons of fish and equivalent to about one-fifth of the global reported catch (Agnew and others 2009).

As noted in the previous *Sunken Billions* report, the resulting inaccuracies in catch statistics are an important source of uncertainty with respect to scientific advice on fisheries management (FAO 2002; Flothmann and others 2012; Kelleher 2002; Pauly and others 2002). Furthermore, the bio-economic calculations used in this report are based on inputs taken from the FAO's FishStat Plus database, which relies on national reporting of catches and therefore does not fully incorporate the considerable amount of catches resulting from IUU fishing.

A potential game changer: Current and future impacts of climate change on fisheries management

Our understanding of the impacts of climate change on fisheries and related ecosystems is constantly improving, and can be organized summarily around several main "vectors"—acidification, sea-level rise, higher water temperatures, and changes in ocean currents. These different vectors, however, are unequally known and hard to model, both in terms of scope—where they will occur, where they will be felt the most—and in terms of severity. For instance, while not as well understood as the other impacts, and more difficult to measure, acidification's impacts are likely to be the most severe and most widespread, essentially throughout any carbon-dependent ecological processes. Likewise, the effects of sea-level change will be felt differently in different parts of the world, including depending on the ecosystems around which it occurs.

Despite this uncertainty, our current state of knowledge is sufficient to understand that these impacts will be felt at two fundamental levels: first on fish stocks, and second, and perhaps more importantly, on the critical marine and coastal ecosystems on which they depend. Climate change is thus becoming a game changer for fisheries management for two reasons: first, it has broadened the necessary scope of action (which, heretofore, had focused disproportionally on the status of stocks) to better include these ecosystems that are at the forefront of the climate change impacts; and, second, it adds a sense of urgency to these needed reforms, because these relatively new and growing impacts come on top of those of overfishing and mismanagement, further increasing the uncertainty level and removing the "safety" that previously allowed depleted stocks to recover post-overexploitation.

The fundamental reforms that are required to help recover the sunken billions must thus follow two parallel and simultaneous paths: (a) stock recovery (giving depleted and overexploited stocks a chance at a comeback, including primarily through capacity reduction and thus the level of effort); and (b) restoring the integrity of the critical habitats on which the stocks depend (including, but not limited to mangroves, coral reefs, and seagrass beds). These two paths will have to adapt over time to climate change's dynamic and changing exigencies.

Notes

1. This report uses, throughout, the term "net benefits" when comparing the relative costs and benefits of fishing activities worldwide. Strict cost-benefit methodology makes a clear distinction between financial analysis and economic analysis. Financial estimates of costs and revenues include all taxes and subsidies, while economic analysis nets them out (since they represent transfer payments, not true costs of production). Therefore, in financial analysis, net costs and revenues are called profits; in economic cost-benefit analysis, net costs and benefits are called net economic benefits.

   In this report, the term "net benefits" is used to indicate that the estimates of "revenues minus costs" represent neither true financial profits nor net economic benefits. They are in-between. There is inadequate knowledge of fishery sector financial costs and revenues to conduct true financial analysis—and similarly, there is inadequate knowledge of fishery sector taxes and subsidies worldwide to conduct true economic analysis. For example, the revenue estimates are only based on one composite unit price times global revenues, and costs are based on estimating operating costs only. Still, the resulting ballpark estimates are very valuable from a policy perspective and to illustrate large-scale trends. Nevertheless, improving the current level of poor fishery sector data is a priority for advancing the fishery reform agenda in the years ahead.

2. FAO data are organized according to this grouping.

3. In addition, many marine fisheries have not been subject to stock assessments.

4. Except if otherwise noted, all tons refer to metric tons.

5. For instance, if we assume that fishing effort production is characterized by a Cobb-Douglas production function of capital (vessel) and labor with a technology coefficient (total factor productivity), reasonable parameters values (for example, exponents on capital and labor equal to 0.5) immediately generates this result.

6. Obtained by discounting the nominal fish price with the U.S. Consumer Price Index (CPI).

7. FAO Fisheries Information and Statistics Branch (FIPS) food balance sheets (FBS) of fish and fishery products.

8. If we assume for the sake of argument that the elasticity of fish price with respect to fish supply is -0.1 (a likely value in the long run [Roy, Tsoa, and Shrank 1991]), fish farming expansion may have reduced the average price of landed fish by some 10 percent.

9. http://www.fao.org/fishery/governance/capture/en.

10. http://www.fao.org/fishery/topic/2014/en.

References

Agnew, D., J. Pearce, G. Pramod, T. Peatman, R. Watson, J. Beddington, and T. Pitcher. 2009. "Estimating the Worldwide Extent of Illegal Fishing." *PLoS ONE* 4 (2): e4570.

Arnason, R. 2007. "Fisheries Management: Basic Principles." In "Fisheries and Aquaculture," edited by Patrick Safran. In *Encyclopedia of Life Support Systems* (EOLSS). Developed under the auspices of UNESCO. Oxford, U.K.: Eolss Publishers. http://www.eolss.net.

———. 2012. "Property Rights in Fisheries: How Much Can ITQs Accomplish?" *Review of Environmental Economics and Policy* 6 (2): 217–36.

Clark, C. 1990. *Mathematical Bioeconomics: The Optimal Management of Renewable Resources.* 2nd ed. New York: John Wiley & Sons.

Costello, C., S. Gaines, and J. Lynham. 2008. "Can Catch Shares Prevent Fisheries Collapse?" *Science* 321 (5896): 1678–81.

Costello, C., J. Lynham, S. E. Lester, S. D. Gaines. 2010. "Economic Incentives and Global Fisheries Sustainability." *Annual Review of Resource Economics* (2): 299–318.

de Fontaubert, C. and I. Lutchman, with D. Downes and C. Deere. 2003. "Achieving Sustainable Fisheries: Implementing the New International Legal Regime." IUCN, Gland, Switzerland, and Cambridge, U.K.

FAO (Food and Agriculture Organization of the UN). 1999. "Fishery and Aquaculture Statistics 1998." FAO, Rome.

———. 2002. *The State of World Fisheries and Aquaculture 2002.* Rome: FAO.

———. 2007a. "Fish and Fishery Products: World Apparent Consumption Statistics Based on Food Balance Sheets (1961–2003)." FAO Fisheries Circular 821, Rev. 8, FAO, Rome.

———. 2007b. "Increasing the Contribution of Small-Scale Fisheries to Poverty Alleviation and Food Security." FAO Fisheries Technical Paper 481, FAO, Rome.

———. 2008. "Climate Change for Fisheries and Aquaculture." Technical Background Document from the Expert Consultation held on April 7–9, FAO, Rome.

———. 2013. Statistical Yearbook. http://www.fao.org/docrep/018/i3107e/i3107e00.htm. FAO, Rome.

———. 2014a. *The State of World Fisheries and Aquaculture 2014 (SOFIA).* Rome: FAO.

———. 2014b. "Fishery and Aquaculture Statistics 2012." FAO, Rome.

———. 2015. "Food Outlook: Biannual Report on Global Food Markets (May)." FAO, Rome. http://www.fao.org/3/a-I4581E.pdf.

———. 2016. *The State of World Fisheries and Aquaculture 2016 (SOFIA).* Rome: FAO.

FAO Fishstat Plus (database). http://www.fao.org/fishery/statistics/software/fishstat/en. FAO, Rome.

Flothmann, S., K. von Kistowski, E. Dolan, E. Lee, F. Meree, and G. Album. 2012. "Closing Loopholes: Getting Illegal Fishing under Control." *Science* 328 (5983): 1235–36.

Gordon, H. S. 1954. "Economic Theory of a Common Property Resource: The Fishery." *The Journal of Political Economy* 62 (2): 124–42.

Hardin, G. 1968. "The Tragedy of the Commons." *Science* 162 (3859): 1243–48.

Kelleher, K. 2002. "The Costs of Monitoring, Control and Surveillance of Fisheries in Developing Countries." FAO Fisheries Circular 976, FAO, Rome.

———. 2005. "Discards in the World's Marine Fisheries. An Update." FAO Fisheries Technical Paper 470, FAO, Rome.

Ostrom, E. 1990. *Governing the Commons: The Evolution of Institutions for Collective Actions.* New York: Cambridge University Press.

Pauly, D., V. Christensen, S. Guénette, Tony J. Pitcher, U. R. Sumaila, C. J. Walters, R. Watson, and D. Zeller. 2002. "Towards Sustainability in World Fisheries." *Nature* 418 (August 8): 689–95.

Roy, N., E. Tsoa, and W. Schrank. 1991. "What Do Statistical Demand Curves Show? A Monte Carlo Study of the Effects of Single Equation Estimation of Groundfish Demand Functions." In *Econometric Modelling of the World Trade in Groundfish*, edited by Schrank and Roy. Dordrecht: Kluwer Academic Publishers.

Schrank, W. E., R. Arnason, and R. Hannesson, eds. 2003. *The Cost of Fisheries Management.* U.K.: Ashgate.

Willmann, R., P. Boonchuwong, and S. Piumsombun. 2003. "Fisheries Management Costs in Thai Marine Fisheries." In *The Cost of Fisheries Management*, edited by W. E. Schrank, R. Arnason, and R. Hannesson, 187–219. U.K.: Ashgate.

World Bank. 2013. *Fish to 2030: Prospects for Fisheries and Aquaculture.* Washington, DC: World Bank.

———. 2014. *Trade in Fishing Services: Emerging Perspectives on Foreign Fishing Arrangements.* Washington, DC: World Bank.

World Bank and FAO (Food and Agriculture Organization). 2009. *The Sunken Billions: The Economic Justification for Fisheries Reform.* Washington, DC: World Bank.

Worm, B., R. Hilborn, J. Baum, T. Branch, and others 2009. "Rebuilding Global Fisheries." *Science* 325 (5940): 578–85.

Basic Approach:
The Bio-Economic Model
and Its Inputs

The bio-economic model used in this study is a slight generalization of the one used in the original *Sunken Billions* study (Arnason 2011; World Bank and FAO 2009). It is a typical aggregate fisheries model and complies with accepted fisheries economics theory and empirical knowledge (see, for example, Anderson [1977]; Anderson and Seijo [2010]; Bjorndal and Munro [2012]; and Clark [1980]).[1]

The model represents a fairly simplified characterization of the global fishery and focuses on the functional interactions between the total global fishing effort level and changes in the total global fish stock. The model is not designed to analyze the performance of individual fisheries or the interactions between different fishing operations within a fishery. It does not predict global seafood market outcomes since it targets the industry's production side and only incorporates demand-side factors to a very limited extent.[2] Nevertheless, its very simplicity allows for the assessment of the economic performance of the global marine fishery in a robust and transparent manner.

Characteristics of the model

The model's approach and steps are summarized as follows:

- Global marine fisheries are treated as one large fishery.
- This fishery is represented in a bio-economic model consistent with fisheries economics theory and supported by empirical knowledge of the global fishery.

- The unknown parameters for the bio-economic model are drawn from best available empirical estimates.
- Empirical information about the state of the global fishery as of the date of the base year is incorporated in the bio-economic model to obtain estimates of net benefits in the base year.
- Application of the bio-economic model results in an evaluation of the maximum sustainable net benefits attainable by the global fishery in the base year.
- The difference between the maximum sustainable economic benefits and the base-year benefits represents foregone economic benefits, referred to as the sunken billions.
- Stochastic simulations are applied to obtain confidence intervals for the value of the sunken billions.
- The bio-economic model is used to evaluate alternative adjustment paths that can lead to the long-run optimal state of global fisheries.

A crucial step in this approach is to model the global fishery as one fishery. This approach presents two main advantages, as follows:

1. The number of data units required to estimate the parameters of the model is relatively small, and the empirical data work is correspondingly reduced.

2. Once finalized, the model is relatively easy to apply. As a result, stochastic simulations—that is, an investigation into the role of different input parameters in generating the results, an examination of different adjustment paths toward the long-run optimal state, studies of impacts of exogenous shocks, and so on—are much easier to conduct than with a collection of disaggregated models. Appendix A looks at the issue of fisheries aggregation in greater detail.

The bio-economic model does not explicitly characterize or incorporate governance and other institutional aspects of the global fishery. Instead, the model focuses on *functional* relationships between fish population biology and human fishing activities. While the level of fishing activities is characterized by the variable "aggregate fishing effort," the model is not concerned with how the fishing effort level is determined. Nonetheless, the model can be applied to produce different outcomes that can be expected under different governance and institutional arrangements. This application is particularly true in terms of the speed or pace at which reforms can be applied. In appendix A, for instance, the model is used to compare how fast fisheries' recovery can be expected depending on the fishery's governance structure, which contributes to the fundamental incentives and constraints facing individual fishers and ultimately to the size of the aggregate fishing effort.

FIGURE 2.1
Maximum sustainable yield and maximum economic yield

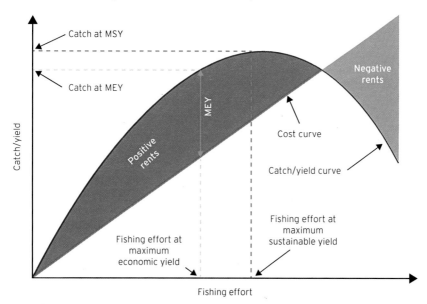

Source: World Bank and FAO 2009.
Note: MEY = maximum economic yield; MSY= maximum sustainable yield.

As illustrated in figure 2.1, the purpose of the aggregate model is to assess the sustainability of the global fishery, and, in this case, to estimate the rent lost to inefficient fisheries management. From a biological standpoint, the level of effort should not exceed that at the maximum sustainable yield (MSY), whereas economically, the positive rent is maximized at the level of catch at maximum economic yield (MEY). Both points are exceeded when a fishery is not managed efficiently, or when the cost curve is artificially lowered, such as when distorting subsidies mask the true cost of fishing, encouraging the fishing effort even when it no longer makes economic sense.

Basic structure of the model

The bio-economic model is dynamic in that it describes the global fishery's evolution over time. Fundamental to this evolution is the fish biomass, which increases or decreases depending on the relative rates of harvest and biological growth. This biological growth is determined in terms of the reproduction net of natural mortality and weight gain of individuals. The time step considered is a year, and the model registers annual changes in its variables.

The model contains eight basic variables, as follows:

- Biomass (total amount of fish resources)
- Biomass growth (gross increase in biomass during a year due to natural processes)
- Harvest (volume of catch)
- Fish price
- Fishing revenues
- Fishing costs
- Net benefits
- Level of fishing effort

The first seven variables are determined internally in the model based on the specified value of the input parameters and the initial value of certain variables. The last variable, fishing effort, is the driver in this model and is exogenous. When it is specified, it acts as the driver of the fishery, generating the seven endogenous variables. It is important to realize that fishing effort is a composite indicator of fishing activity and includes the number, type, and power of fishing vessels and the type and amount of fishing gear. As a result, quantifying fishing effort in even a single fishery is difficult, and there is considerable uncertainty about the global level of effort (Pauly and Zeller 2016; World Bank and FAO 2009).

The different relationships between these variables are characterized by five core functions:

- A net biomass growth as a function of fish stock biomass natural growth minus harvest
- A harvesting function determining the catch volume
- A function describing the total cost of fishing
- A price function characterizing how the price of catch is determined
- A benefit function that calculates the net benefits[3]

Figure 2.2 describes the bio-economic model's basic structure and operation. The arrows indicate the direction of causality, that is, how the levels of the variables are determined internally in the model. At the beginning of each year, the biomass starts at a certain level. With the fishing effort level specified, fishing effort and the biomass level combine to determine the harvest level as specified by the harvesting function. The harvest level and the natural biomass growth, in turn, combine to determine the biomass level at the beginning of the next period. The process is repeated each year, and the fish population evolves over time.

On the economics side, specifying the fishing effort level leads to the determination of fishing costs according to the cost function. Biomass level determines

FIGURE 2.2
Structure of the bio-economic model

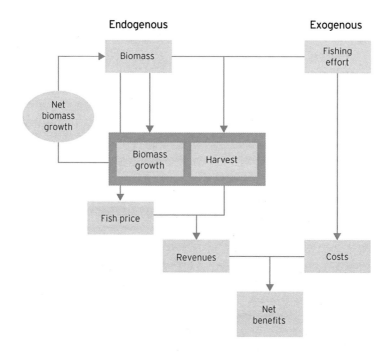

the catch price according to the price function. Multiplying the harvest with the fish price generates total fishing revenues. Subtracting fishing costs from the revenues produces the net benefits from the fishery during this period.

Core functions of the model

The actual equations for the five core functions used in the model are as follows:

(2.1)	$\dot{x} = \alpha \cdot x - \beta \cdot x^{\gamma} - y$	(Net biomass growth function):
(2.2)	$y = q \cdot e \cdot x^{b}$	(Harvesting function),
(2.3)	$C = c \cdot e + fk$	(Fishing cost function),
(2.4)	$p = a \cdot x^{d}$	(Fish price function),
(2.5)	$\pi = p \cdot y - C$	(Benefit function).

The variable x represents the level of biomass, $\dot{x}$ its net change within a given year, y harvest, e fishing effort, C fishing costs, p the fish price, and π net benefits. The symbols α, β, γ, q, b, c, fk, a, and d are model parameters.

The following section focuses on the basic logic that justifies the construct of each function. Appendix B includes the rationale behind the selection of the specific functional forms, as well as the properties of these functions.

Net biomass growth function

Equation (2.1) describes the evolution of fish stock biomass as the natural growth of biomass (biological processes of reproduction net of natural mortality and weight gain) minus the fishing mortality or harvest. As is typically done in the modeling of population dynamics with unregulated reproduction, the natural biomass growth is modeled as a function of the biomass itself (x). The specific expression of the first part of the equation (2.1), $\alpha \cdot x - \beta \cdot x^\gamma$, is called the Pella-Tomlinson biomass growth function. While the parameter α represents the intrinsic growth rate of the biomass, different combinations of the three parameters α, β, and γ characterize the carrying capacity, maximum sustainable yield, and other important reference points of the population (see appendix B, table B.1).

The use of the Pella-Tomlinson functional form represents an improvement over the original *Sunken Billions* study, where logistic and Fox biomass growth functions were used. This is an improvement because the Pella-Tomlinson function is more flexible than the other two, and in fact incorporates both as special cases. If $\gamma = 2$, the Pella-Tomlinson function becomes the logistic function, and as γ approaches unity, it converges to the Fox biomass growth function. If γ is between unity and two, other biomass growth functions are defined. See appendix B for further details.

Harvesting function

The harvesting function in equation (2.2) describes the volume of global catch as a function of the aggregate fishing effort (e) and biomass (x). In this equation, the parameters q and b are both positive, implying that a larger harvest is achieved with a higher level of fishing effort and with a larger fish biomass. The coefficient q is often referred to as the catchability coefficient, describing the fishing efficiency. The coefficient b indicates the degree of schooling behavior by the fish, which also affects fishing efficiency. The implications of the schooling parameter will be discussed further.

Fishing cost function

In this model, the cost function refers only to the actual cost of catching the fish, and does not take into account other related costs, such as that of fisheries management, enforcement, and so on, which generally are not borne by the fishers themselves.

In equation (2.3), the aggregate fishing cost is thus described in two parts: the unit variable cost that is multiplied by increases with the fishing effort level (e)

and the fixed cost (fk) that is independent of the fishing effort level. For modeling purposes, the value of the fixed cost is assumed to be zero: instead annualized fixed costs are incorporated in the variable cost. While separating variable and fixed costs is important in financial accounting, it is not important for a long-term economic model. While investment costs may appear fixed to the fishing operator, they are not fixed from a long-term macro perspective.

Fish price function

Equation (2.4) describes the catch price as a function of biomass level (x). The idea behind the modeling of fish price as dependent on the biomass is to reflect the biological reality that, as fish stocks increase, landings increasingly consist of more valuable species and larger individuals, which typically command higher prices (Herrmann 1996; Homans and Wilen 2005). On this basis, global fish biomass recovery is also expected to increase the average price of all landed catches. The parameter d, referred to as an elasticity of price with respect to biomass, describes the strength of correlation between the biomass and average fish price of marine catches. This parameter is discussed in detail in appendix B.

Note that the fish price in this model does not represent market price as determined by supply and demand, but rather, considers the demand-side conditions as fixed when predicting increases or decreases in fish price.[4]

Benefit function

Equation (2.5) simply states how the net benefits are obtained in the model: aggregate revenues, defined as a product of the average unit price (p) and the catch level (y), minus aggregate costs (C). The real challenge lies in determining the actual level of net benefits that the global fishery achieves. Formally, the net benefits from fishing are defined as the social value of landed catch less the social value of the inputs used to produce the catch. In economic theory, the social value of anything, including inputs and outputs, equals the quantity in question multiplied by the "true" price (Debreu 1959). Since taxes and subsidies are transfer payments, not true costs, they are netted out in calculating net benefits.Unfortunately, the current state of fishery sector data does not allow an accurate calculation of either global financial benefits (that is, profits), or net economic benefits.[5]

Estimation of model inputs

As seen in equations (2.1)–(2.5), the bio-economic model incorporates a number of parameters, whose numerical values must be estimated. The model also requires initial (base-year) values for certain variables. Table 2.1 lists the value of these model inputs used to generate results for the 2012 base year. Note that a relatively small number of model inputs drive the estimates of the total

TABLE 2.1
Inputs for bio-economic model for the 2012 base year

Input	Symbol	Unit	Value
Maximum sustainable yield	MSY	Million tons	102.0
Biomass carrying capacity	Xmax	Million tons	980.0
Pella-Tomlinson exponent	γ	n.a.	1.188
Schooling parameter	b	n.a.	0.71
Elasticity of price w.r.t. biomass	d	n.a.	0.22
Net biomass growth in 2012	$\dot{x}$	Million tons/year	8.0
Landed volume in 2012	y	Million tons	79.7
Landings price in 2012	p	US$/kg	1.26
Net benefits in 2012	π	US$ billion	3.0

Sources: various (see text).
Note: n.a. = not applicable; w.r.t. = with respect to. Some of the model parameters that appear in equations (2.1)–(2.5) are absent in the table because they are internally calculated based on the values of other parameters and the base-year value of certain variables. More specifically, parameter values for α and β in equation (2.1) are derived using the values of MSY, Xmax, and the Pella-Tomlinson exponent (γ). The value of q in equation (2.2) is determined using the base-year values of e, x, and y. Similarly, the value of a in equation (2.4) is given by the base-year value of x and p. The value of c in equation (2.3) is determined through the base-year value of e, p, y, and π, and is estimated in the model. The initial volume of biomass (x) is obtained using equation (2.1) and the base-year estimate of the biomass growth ($\dot{x}$). The base-year fishing effort (e) is normalized at unity. The formulas used for the derivations of the various parameters are found in table C.4 in appendix C.

sunken billions, highlighting the importance of the accuracy of their numerical values. Consequently, this study gives serious consideration to the uncertainty surrounding the value of model inputs and the potential errors the model outputs are subject to, as will be discussed further. The rest of this section briefly explains the role of these model inputs and how these values are obtained, with additional discussions on model inputs provided in appendixes B and C.

The justification for the selection of model inputs is explained in greater detail in appendix B, but the main choices can be summed up as follows:

- To estimate the MSY, the input selected took into account recent empirical estimates and a number of other indications. The value of 102 million tons used in this report, which represents the total global fisheries, is an upward revision from the "conservative" 95 million tons' estimate adopted in the initial *Sunken Billions* study of 2009.

- The carrying capacity (Xmax) was assumed at 9.6 times the MSY, and Xmax is therefore set at 980 million tons. This parameter is rather difficult to estimate, and it varies between 5 and 15 times the MSY. Based on the sensitivity analysis, however, this parameter does not substantially influence the ultimate results.

- The Pella-Tomlinson exponent determines the skewness of the Pella-Tomlinson biomass growth function in equation (2.1) (see appendix B), and it is assumed to equal 1.188, based on empirical estimates. Based on a database of landings and biomass for 147 fish stocks covering the main types of commercial fish stocks fairly widely distributed around the world, Thorson and others (2012) found that the MSY occurred approximately at 40 percent of the carrying capacity of these stocks; this corresponds to a Pella-Tomlinson exponent, γ, approximately equal to 1.188.

- The schooling parameter (b) characterizes the schooling behavior of the stocks, which in turn affects fishing efficiency. The parameter normally has a value between zero and unity. The lower the schooling parameter, the more pronounced the schooling behavior is, and the less dependent on biomass is the harvest. In this report the long-term average of 0.72 was adopted.

- The price elasticity with respect to biomass (d) used here is 0.22, which means that, if the global biomass doubles (a 100 percent increase), then the average price of landings increases by 22 percent. The parameter chosen here is based on the evidence suggesting that, in the short term, the price elasticity to the introduction of improved management systems varies between 10 to 30 percent.

- The biomass growth ($\dot{x}$) in the base year assumed here is 8 million tons because, although the evolution of fish stocks is far from uniform across the world and declines undoubtedly prevail in some regions, it appears that, in 2012, global fish stocks may actually have increased by that amount.

- The volume of landings (y) in the base year was estimated on 79.7 million tons, based on the FAO FishStat Plus estimates.

- The landed catch price (p) in the base year was set at $1,260 per ton, based on the analysis of over 4,000 individual fisheries (Costello and others 2012, 2015).

Finally, the global fisheries net benefits are an important input into the global bio-economic model. They measure the benefits generated by the fishing activity, and, as such, do not necessarily coincide with accounting profits. (See appendix A for further discussion of the concept of net financial versus economic benefits.) Any estimation of the profitability of the global fishing fleet suffers from a scarcity of reliable data on cost and earnings, as in most fishing nations, fisheries cost and earnings statistics are not systematically collected. Profitability data are particularly deficient for small-scale, artisanal, and subsistence fishing fisheries, which collectively compose a large proportion of the value of the global marine fishery. It should also be noted that, even when profitability data are collected, fishers are often reluctant to provide complete and accurate information, so that the data obtained are seldom very reliable. Nevertheless, base-year net benefits were estimated from a range of empirical data pertaining to the year 2012, and valued at US$3.0 billion.

Notes

1. Appendixes A and B provide a more detailed explanation and specifications of the model.
2. World Bank (2013), for example, provides analysis of the global market for fish and fish products.

3. See the next section for the equations.

4. Refinement of the modeling of fish price trends, taking into account such factors as demand-side considerations and the impact of supply-side farmed fish, is warranted, but would require separate studies to be done.

5. See footnote 1 in chapter 1 for an explanation of the use of the term "net benefits," and the difference between financial and economic analysis.

References

Anderson, L. 1977. "The Economics of Fisheries Management." Johns Hopkins University Press, Baltimore.

Anderson, L., and J. Seijo. 2010. *Bioeconomics of Fisheries Management.* New York: John Wiley & Sons.

Arnason, R. 2011. "Loss of Economic Rents in the Global Fishery." *Journal of Bioeconomics* 13: 213–32.

Bjorndal, T., and G. Munro. 2012. *The Economics and Management of World Fisheries.* New York: Oxford University Press.

Clark, C. W. 1980. "Towards a Predictive Model for the Regulation of Fisheries." *Canadian Journal of Fisheries and Aquatic Science* 37: 1111–129.

———. 1990. *Mathematical Bioeconomics: The Optimal Management of Renewable Resources.* 2nd ed. New York: John Wiley & Sons.

Costello, C., D. Ovando, T. Clavelle, C. Strauss, R. Hilborn, M. Melnychuk, T. Branch, S. Gaines, C. Szuwalski, R. Cabral, D. Rader, and A. Leland. 2015. "Have Your Fish and Eat Them Too." Unpublished manuscript.

Costello, C., D. Ovando, R. Hilborn, S. Gaines, O. Deschenes, and S. Lester. 2012. "Status and Solutions for the World's Unassessed Fisheries." *Science* 338: 517–20.

Debreu, G. 1959. "Theory of Value." Cowles Foundation, Monograph 17, Yale University Press, New Haven.

Herrmann, M. 1996. "Estimating the Induced Price Increase for Canadian Pacific Halibut with the Introduction of Individual Vessel Quota System." *Canadian Journal of Agricultural Economics* 44: 151–64.

Homans, F., and J. Wilen. 2005. "Markets and Rent Dissipation in Regulated Open Access Fisheries." *Journal of Environmental Economics and Management* 49: 381–404.

Pauly, D., and D. Zeller. 2016. "Catch Reconstructions Reveal that Global Marine Fisheries Catches Are Higher Than Reported and Declining." *Nature Communications* 7, Article number: 10244. doi:10.1038/ncomms10244.

Thorson, J., J. Cope, T. Branch, and O. Jensen. 2012. "Spawning Biomass Reference Points for Exploited Marine Fishes, Incorporating Taxonomic and Body Size Information." *Canadian Journal of Fisheries and Aquatic Science* 69: 1556–68.

World Bank. 2013. *Fish to 2030: Prospects for Fisheries and Aquaculture.* Washington, DC: World Bank.

World Bank and FAO (Food and Agriculture Organization). 2009. *The Sunken Billions: The Economic Justification for Fisheries Reform.* Washington, DC: World Bank.

The Sunken Billions: Main Results

Following the steps in chapter 2, the bio-economic model, operationalized with updated estimated model inputs, was used to generate estimates of the foregone benefits of the global fishery for 2012. The total net economic gain of adopting more sustainable fisheries management, which we call the sunken billions, are estimated at US$83 billion for that year. Conversely, this finding indicates that the world's currently unsustainable fisheries management practices have led to globally depleted fish stocks that produce $83 billion less in annual net benefits than would otherwise be the case.

This result affirms the basic conclusion of the 2009 report, namely that adopting more sustainable fishing practices would pay for themselves many times over, and greatly improve the livelihoods of millions of people. In addition, the model's breakdown by regions shows that the sustainability of current fishing practices, and hence their economic returns, varies considerably across regions. In Asia and Africa, most fisheries appear to be vastly overexploited, but in Oceania the total catch is likely *below* the maximum sustainable yield (MSY) level. While subject to greater uncertainty than the global estimates, the regional analysis suggests that adopting more sustainable fishing practices would benefit Asian and African countries, including those like China that possess large fishing fleets, the most.

Main results

Table 3.1 summarizes the model's key numerical outputs. Given the initial conditions set in 2012, the maximum net benefits attainable on a sustainable basis from the global fishery are estimated to be $86.3 billion annually (column (2)). In comparison, and as chapter 2 highlighted, the 2012 estimated net benefits were $3 billion (column (1)). As a result, the estimated foregone net benefits, or what we called the 2012 sunken billions in the global marine fishery were $83.3 billion (column (2) – (1)) (all of these values are expressed in 2012 US dollars).

Table 3.1 includes several other important conclusions:

- First, the bio-economic model estimates that to achieve the maximum net benefits, the aggregate fishing effort would need to be reduced by about 44 percent relative to the fishing effort that prevailed in 2012. Since the fishing cost is assumed to be linearly related to the fishing effort and no fixed costs are considered in this study, the total fishing cost would also be lower by 44 percent in the optimal state.

- Second, by keeping the fishing effort at this lower level, the global biomass of commercially exploited fish stocks would reach about 580 million tons, which is 2.7 times as high as the 2012 estimated biomass of 215 million tons.

- Third, the estimated sustainable harvest in the optimal state, or maximum economic yield (MEY), is 89.7 million tons. Thus, and despite the substantially lower fishing effort (–44 percent), and since the biomass is larger, the estimated MEY is about 10 million tons higher than the 2012 amount harvested. The model reveals the remarkable benefit that could be derived from restoring fish

TABLE 3.1
Summary results of the bio-economic model for the 2012 base year

Variable	Symbol	Unit	(1) 2012 Baseline	(2) Sustainable optimal	(2) - (1) Difference	(2)/(1) Ratio: optimal scenario/ 2012 baseline
Biomass	x	Million tons	214.9	578.6	363.7	2.692
Harvest	y	Million tons	79.7	89.7	10.0	1.126
Effort	e	n.a.	1.000	0.557	-0.443	0.557
Landings price	p	US$/kg	1.26	1.567	0.31	1.244
Revenues	$p \cdot y$	US$ billion	100.422	140.6	40.2	1.400
Costs	C	US$ billion	97.422	54.3	-43.1	0.557
Net benefits	π	US$ billion	3.0	86.3	83.3	28.767
Net benefits per unit effort	π/e	n.a.	3.0	154.9	151.9	51.645

Source: Model output.
Note: n.a. = not applicable.

stocks to a healthy level: higher sustainable harvests could be achieved with far less fishing effort.[1]

- Fourth, the estimated average landings price (the ex-vessel price of catch) would be 24.4 percent higher in the optimal state of the fishery than it was in 2012, due to the proportionately higher share of high-value species and large individual fish in the landed catch.

- Fifth, the combination of much lower fishing cost and higher harvest would lead to an almost 30-fold increase in the net benefits in the optimal state over that which was achieved in 2012 ($86.3 billion versus $3 billion).

- Finally, and given the substantial reduction in the total fishing effort in the optimal state (fewer boats and larger fish stocks), the difference in estimated net benefit for each unit of effort is even greater between the optimal state and the state in 2012. According to the model results, the net benefits per unit effort in the optimal state of the global fishery would be over 50 times higher than they were in 2012. This, in turn, means that fishing activities would be much more profitable in the optimal state than they are now.

To appreciate the contribution of each of these factors to the gains in net benefits under the optimal sustainable state, figure 3.1 illustrates the breakdown of each of these gains for 2012. The $83.3 billion of additional benefits under the optimal state can be attributed to three factors, as follows: (i) higher sustainable harvest attainable due to larger fish biomass; (ii) lower fishing costs due to lower fishing effort; and, (iii) higher unit prices of landings due to improved species composition of the global stock. As seen in figure 3.1, over half of the increased economic gains in the optimal equilibrium can be attributed to cost reductions

FIGURE 3.1

Breakdown of 2012 sunken billions estimate: Sources of additional economic benefits in the optimal sustainable state

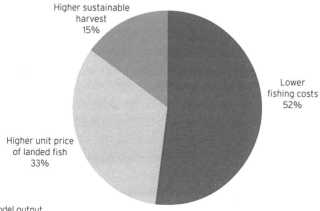

Source: Model output.

due to reduced fishing effort. About one-third comes from higher landings price and the remaining 15 percent stems from increased harvest following the recovery of overexploited fish stocks. Consequently, the majority (85 percent) of the potential gains are the result of improved economic returns from fishing rather than any increases in the catch itself. This result also reinforces the message from the original report that the problem facing the global marine fisheries is an economic one as much as it is a biological problem (World Bank and FAO 2009).

The reduction in fishing effort is the fundamental driver of the gains under the optimal state. A lower fishing effort leads to higher biomass and these two changes make the three types of gains possible.[2]

Sensitivity analysis and confidence intervals

As previously noted, the various model outputs presented in this chapter are subject to considerable uncertainty. The estimates of the maximum attainable net benefits and the foregone benefits depend on the bio-economic model employed to describe the global fishery as well as the data inputs. In addition, the bio-economic model is an approximation of the real world and is subject to errors. The specific values of input parameters used in this study represent best estimates given available data and knowledge, but the quality of fisheries data, and therefore model inputs, are often highly uncertain (see chapter 2).

This section presents two sets of effort to come to an assessment of the extent of possible errors in model predictions. First, a sensitivity analysis is conducted to analyze how model outputs change in response to changes in the value of input parameters. Second, statistical procedures are employed to generate confidence intervals, which indicate a range in which the true value of sunken billions is considered to fall at a specified level of confidence.

Sensitivity analysis

If the specification of the bio-economic model is accepted as given, the estimates of the foregone economic benefits in global marine fisheries and all the intermediate variables depend entirely on the input parameter values supplied to the model. To that end, the authors conducted a sensitivity analysis to assess the sensitivity of the model outputs to the various model inputs.

Fixed costs are set to zero for theoretical reasons (see chapter 2), and this value therefore is not subject to uncertainty. As shown in table 2.1, there are nine others exogenous inputs for this model, for which the sensitivity analysis is thus conducted for the remaining nine model inputs. Figure 3.2 graphically depicts the results for the sunken billions estimate. In each part of this figure, the horizontal axis represents the deviations from the adopted values of model inputs (as in table 2.1) of up to 50 percent in both negative and positive directions. The corresponding estimates of foregone economic benefits are measured along the

FIGURE 3.2

Sensitivity of estimated foregone benefits to the model inputs

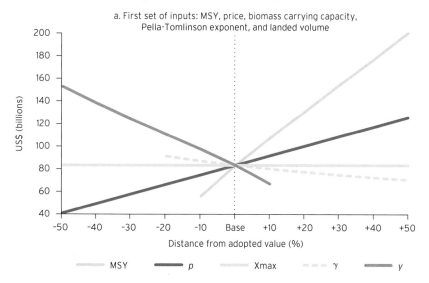

a. First set of inputs: MSY, price, biomass carrying capacity, Pella-Tomlinson exponent, and landed volume

US$ (billions)

Distance from adopted value (%)

MSY   p   Xmax   γ   y

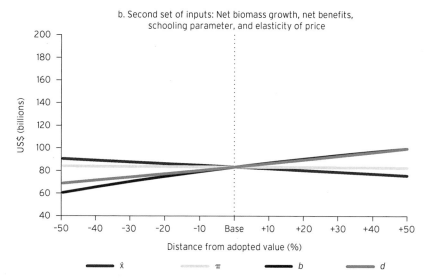

b. Second set of inputs: Net biomass growth, net benefits, schooling parameter, and elasticity of price

US$ (billions)

Distance from adopted value (%)

ẋ   π   b   d

Source: Model output.
Note: MSY = maximum sustainable yield; p = landings price; Xmax = biomass carrying capacity; γ = Pella-Tomlinson exponent; y = harvest. ẋ = biomass growth; π = net benefits; b = schooling parameter; d = elasticity of price with respect to biomass.

vertical axis. The relationship between the specific model input values and the resulting foregone net benefits is drawn as curves. The steeper the slope of these curves, the greater is the numerical sensitivity of the foregone benefits estimates to that particular input parameter.

TABLE 3.2
Confidence intervals for foregone net benefits

Degree of confidence (%)	Confidence intervals (US$ billions)
80	[65.1, 99.0]
90	[57.2, 102.5]
95	[49.7, 104.9]

Source: Model output.

The sensitivity analysis summarized in both parts of figure 3.2 clearly demonstrates that the estimated foregone net benefits are most sensitive to the MSY level, followed by the base-year volume of harvest (y), the base-year landings price (p), the elasticity of the landings price with respect to biomass (d), and the schooling parameter (b). By contrast, the estimated foregone benefits virtually do not change with the level of biomass carrying capacity (Xmax) and are almost invariant to the value of base-year profits (π).

Stochastic analysis and confidence intervals

One way to account for the uncertainty concerning the global bio-economic model and the inputs entered is to regard them as stochastic and subject to probability distributions. With specifications of these probability distributions, it is possible to obtain probability distributions and confidence intervals for the model outputs by employing Monte Carlo stochastic simulations (Davidson and MacKinnon 1993).

The stochastic analysis of this study is limited to addressing the uncertainty of the model inputs, while the uncertainty about the appropriateness of the bio-economic model itself is not accounted for. Of the 10 model inputs to the bio-economic model, nine were regarded as stochastic, as the value of fixed costs of fishing is set to zero and regarded as nonstochastic.

Stochastic outcomes for the 2012 foregone net benefits were generated by drawing from the probability distributions specified for the input parameters. The results are presented in terms of confidence intervals for the value of foregone economic benefits. The previous comparison between the 2012 estimated net benefits and the estimated maximum attainable benefits resulted in the estimated foregone benefits of $83.3 billion. This calculation used point estimates of the mean of the benefit estimates. Table 3.2 shows the confidence intervals for the true value of the foregone economic benefits, according to the stochastic simulations. According to the stochastic specifications of this study, a 95 percent confidence interval for the true foregone benefits is estimated to be between $49.7 billion and $104.9 billion.

Regional analysis of the estimated sunken billions

This section provides a summary assessment of the foregone net benefits in the fisheries of the world's main regions. This study uses five regions, as follows: Asia;

the Americas (including Central, North and South America); Africa; Europe; and Oceania (including Australia, New Guinea, New Zealand, and the Pacific Island States). The results by region are much less reliable than the global results, mainly because fisheries data by region are less dependable than the global data. Regional boundaries are imprecise from a fisheries perspective, and the regional estimates have not been subjected to the same amount of scrutiny.

First, much of the United Nations' Food and Agriculture Organization (FAO) fisheries data, on which this study depends quite heavily, are not organized on a regional basis. This applies, for instance, to data on the state of fish stocks and fish prices. Second, the process of aggregating regional data to the global level tends to smooth out the errors in regional data. Finally, it is inherently problematic to classify fisheries data by region: many fish stocks straddle or migrate across regional boundaries; fishing takes place where the fish are found, while fishing vessels can travel across regional boundaries; and the catch landings, on which much of the standard fishery statistics are based, may take place in different regions than the actual fishing.

Since the assessment of regional fisheries is a more recent extension of the sunken billions project, the results have been subject to less peer review and scrutiny compared to the global results. That caveat also suggests that the regional findings have a higher uncertainty than the global findings.

Model inputs

The same bio-economic model is used for the regional fisheries as was for the aggregate global fishery, once again using 2012 as the base year. Table 3.3 summarizes the model inputs for the five regions, as well as those from table 2.1 for the global fishery. In preparing regional inputs, the weighted averages across regions for the landings price and schooling parameter were checked to ensure that they matched the global values, while the sum of all regional values adds up to the corresponding global values for the other variables.

The regional MSY values are estimated in the same way as for the global fishery. The long-term history of landings is considered in combination with a broad assessment of the degree of fish stock overexploitation to reach a preliminary MSY estimate in each region. This estimate is then modified in light of other available information, including biological estimates of MSY by major species groups and maximum regional yield estimates. The regional estimates clearly highlight the concentration of commercial marine fishing in Asia, the Americas, and Europe.

In contrast, not much is known about the actual carrying capacity of most fish stocks that are subject to commercial exploitation. To determine the regional carrying capacity, the global Xmax/MSY ratio that was adopted in chapter 2 is used for the regions, with certain modifications to reflect regional differences in biological productivity.

TABLE 3.3
Inputs for bio-economic model for the 2012 base year, by region

Inputs	Symbol	Unit	Asia	Americas	Africa	Europe	Oceania	Global
Maximum sustainable yield	MSY	Million tons	50	23	8	18	3	102.0
Biomass carrying capacity	Xmax	Million tons	450	225	72	205	28	980.0
Pella-Tomlinson exponent	γ	n.a.	1.188	1.188	1.188	1.188	1.188	1.188
Schooling parameter	b	n.a.	0.72	0.66	0.70	0.75	0.69	0.71
Elasticity of price w.r.t. biomass	d	n.a.	0.22	0.22	0.22	0.22	0.22	0.22
Net biomass growth in 2012	$\dot{x}$	Million tons/ year	-1.4	5.0	-0.3	3.8	0.9	8.0
Landed volume in 2012	y	Million tons	41.5	17.9	5.5	12.7	2.1	79.7
Landings price in 2012	p	US$/kg	1.44	0.81	1.40	1.25	1.15	1.26
Net benefits in 2012	π	US$ billion	-2.2	2.9	0.2	1.7	0.4	3.0

Sources: Various (see text).
Note: n.a. = not applicable; w.r.t.= with respect to.

On the same arguments as presented in chapter 2, and in the absence of alternative information, a Pella-Tomlinson exponent of 1.188 is adopted for all five regions.

The regional schooling parameter values reflect the very different mix of fish stocks in various fisheries across the five regions. Small pelagic species, which typically have a strong schooling tendency, account for a particularly high percentage of the fisheries in the Americas and Oceania. This result suggests smaller schooling parameters for these regions and correspondingly larger ones for the others.

Since much of the fish harvest is in effect a global commodity, the same elasticity of price with respect to biomass is adopted for all five regions. Note, however, that since the degree of stock overexploitation differs across regions, the impact of improved fisheries on fish prices will differ accordingly.

The 2012 growth of global marine fisheries biomass was estimated at 8 million tons. Many indicators point to most of this biomass growth occurring in Europe and the Americas, where this growth was estimated to be 3.8 million and 5 million, respectively, while, in contrast, stocks in Asia and Africa continued to decline, by 1.4 million and 0.3 million, respectively. The biomass growth estimates adopted for the regions reflect this phenomenon. On the other hand, it is estimated that the largest biomass growth relative to the biomass carrying capacity occurred in Oceania, where the 2012 estimated biomass growth was 3.2 percent of the estimated carrying capacity. The volume of fish biomass that Asia lost in 2012 amount to 0.3 percent of its carrying capacity, while Africa lost biomass that is 0.4 percent of its carrying capacity. The corresponding global biomass gain is 0.8% of the carrying capacity.

As is the case with the global figure, the 2012 landings volume by region draws on the FAO FishStat Plus database. Again, much of marine catches are concentrated in Asia (52 percent), the Americas (22 percent), and Europe (16 percent).

The landed catch unit price is calculated for each region based on the estimated landings price by species group at the global level, and the species composition of landings by region drawn from the FAO FishStat Plus database. Since the species composition of landings is quite different across the various regions, the average price of landed catch between regions also sees considerable differences. In particular, a large proportion of the landings in the Americas and, to a lesser extent, Oceania, comprise relatively low-value pelagic species, while catches in Africa and, in particular, Asia, include a high proportion of comparatively high-value species. As a result, the estimated average price of catch there is higher than the global average, with $1.44/kg in Asia and $1.40/kg in Africa, compared to a global average of $1.26/kg. It should be noted that these two regions are also where the biomass is estimated to be declining.

For the same reason as for global fisheries, fixed costs are set to zero for regional fisheries.

From the standpoint of net benefits, much evidence points to comparatively better results in some of the fisheries in the Americas as well as in Northern Europe (North Atlantic), and in the large pelagic fisheries of Oceania. On the other hand, Asian and African fisheries, taken as a whole, do not appear to be very profitable. As was the case for global estimates, the table 3.3 regional profitability figures include subsidies, which are comparatively high in Europe and Asia (Sumaila and others 2013). African fisheries achieved a level of profitability equivalent to that of the global average, with net benefits at 3 percent of the revenues. In contrast, regional fisheries in Asia operated at a loss of US$ 2.2 billion which, given its dominant share in global fisheries (more than half of global catches), reduces the global fisheries profitability to some 3 percent of revenues.

Main results

The presentation of regional results begins with the estimated biological state of the fish stocks in each region. Table 3.4 shows the 2012 estimated biomass, the estimated biomass that supports MSY, and the difference between the two. The results indicate that the degree of biological overexploitation differs widely between the five regions, with a lower percentage indicating a higher level of

TABLE 3.4

Estimated biomass in 2012 and biomass at maximum sustainable yield

	Biomass in 2012 (x_0)	Biomass corresponding to MSY(XMSY)	x_0/XMSY
Asia	84.8	180.0	0.47
Americas	82.4	90.0	0.90
Europe	52.9	82.0	0.65
Africa	9.3	28.8	0.32
Oceania	11.2	11.1	1.01
Total[a]	240.6	391.9	0.61
Global[b]	214.9	392.2	0.55

Source: Model output.
Note: MSY = maximum sustainable yield.
a. Total of all five regions.
b. Global results from table 3.1.

biological overexploitation. In Oceania, the biomass of commercially exploited fish stocks is estimated to be very close to or even slightly above the biomass level corresponding to the MSY. In the Americas, the total commercial biomass is not much below the MSY level, with the 2012 estimated biomass reaching 90 percent of the estimated MSY biomass. In Europe, that figure is 65 percent, with lower figures of 47 percent in Asian fisheries and 32 percent in the Africa region.

As these figures clearly indicate, the change that would be needed to bring fisheries close to a state that supports the MSY, and then the long-run optimal equilibrium, is greatest for Asia and Africa, less so for Europe, and more modest for Oceania and the Americas.

The remaining regional results are presented in terms of the difference between the optimal equilibrium and the current situation in 2012 (table 3.5). Table 3.1 presents the equivalent global results in the column labeled "Difference." In table 3.5, the figures in the row labeled "Net benefits" represent the regional estimates of sunken billions.

These results show that in the fisheries of all five regions there is considerable economic waste in terms of excess fishing effort compared to the optimal level. However, the degree of economic waste differs widely between the regions. The level of effort in excess of what would be optimal in the long run ranges from 19 percent in Oceania, to 26 percent in the Americas, 34 percent in Europe, 51 percent in Africa, and 52 percent in Asia.

In table 3.5, it should be noted that the sum of the estimated foregone net benefits for all five regions amounts to $84.6 billion, which is slightly higher than the estimate of $83.3 billion for the global fishery regarded as one fishery (table 3.1), but this difference is statistically insignificant. Furthermore, this difference can be explained, as follows: in the global analysis, the same level of effort is applied to the global fishery, as a whole, whereas in the regional breakdown, the level of effort can be adjusted for each specific region.

TABLE 3.5
Difference between the optimal sustainable state and the current state of fishery, by region

Variable	Symbol	Unit	Asia	Americas	Europe	Africa	Oceania	Total[a]
Biomass	x	million tons	176.6	59.7	69.7	29.8	7.1	342.9
Harvest	y	million tons	3.0	1.1	3.0	1.9	0.3	9.3
Effort	e	n.a.	-0.52	-0.26	-0.34	-0.51	-0.19	n.a.
Landings price	p	US$/kg	0.40	0.10	0.25	0.52	0.13	n.a.
Revenues	$p \cdot y$	US$ billion	22.4	2.8	7.7	6.6	0.6	40.2
Costs	C	US$ billion	-32.4	-3.0	-4.9	-3.8	-0.4	-44.4
Net benefits	π	US$ billion	54.8	5.9	12.5	10.4	1.0	84.6

Source: Model output.
Note: n.a. = not applicable.
a. Total of all five regions.

As these regional results show, at $54.8 billion the estimated foregone net benefits are by far the largest in Asia. In fact, according to these estimates, Asia accounts for almost two-thirds of the total foregone economic benefits in the global fishery. Figure 3.3 highlights the relative contribution of each of the regions to the total sunken billions estimate.

To a certain extent, the relative share of Asia in the sunken billions can be explained by the sheer size of Asian fisheries. They are by far the largest in the world, and account for more than half of the global marine fisheries production, but according to the estimates of this study, they are also the least efficient.

FIGURE 3.3
Regional distribution of total sunken billions

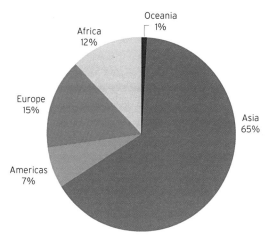

Source: Model output.

Once again, the caveats on the quality and the reliability of the regional estimates of the sunken billions must be emphasized. This, in turn, calls for continued efforts to report, compile, and analyze fisheries statistics from individual countries accurately to improve regional estimates.

Notes

1. However, since the 2012 catches were comparatively low, the harvest gain in the optimal state compared to recent average years would be less than 10 million tons.
2. The individual contributions of the lower fishing effort and higher biomass cannot be truly separated. The problem is one of attributing outputs to individual inputs. On the other hand, it is possible to allocate the gains in overall net benefits to the three factors: cost reduction, fish price increase, and greater harvests. Net benefits are expressed as $\pi = p \cdot y - c$, where p indicates price, y harvest, and c costs. Therefore, the change in net economic gains from some initial time t_0 to some new equilibrium at time t^* is $\Delta \pi = (p(t^*) - p(t_0)) \cdot y(t_0) + (y(t^*) - y(t_0)) \cdot p(t_0) - (c(t^*) - c(t_0))$, where Δ is the difference operator so that $\Delta \pi = \pi(t^*) - \pi(t_0)$, etc.

References

Davidson, R., and J. MacKinnon. 1993. *Estimation and Inference in Econometrics.* U.K.: Oxford University Press.

FAO Fishstat Plus (database). http://www.fao.org/fishery/statistics/software/fishstat/en. FAO, Rome.

Sumaila, R., V. Lam, F. Le Manach, W. Schwartz, and D. Pauly. 2013. "Global Fisheries Subsidies. Note." IP/B/PECH/IC/2013-146, European Parliament, Brussels.

World Bank and FAO (Food and Agriculture Organization). 2009. *The Sunken Billions: The Economic Justification for Fisheries Reform.* Washington, DC: World Bank.

Dynamics of Global Fisheries Reform: Recovering the Sunken Billions

Evolution of global fisheries between 2004 and 2012

Chapter 3 estimated the cost of the sunken billions for 2012 based on modeling of a more sustainable and optimal situation. This chapter looks, first, at the dynamics of change relative to the previous sunken billions analysis for 2004, and second, at the dynamics of hypothetical future pathways to the optimal state of global fisheries.

A standard technique to compare the estimated actual and optimal fisheries states of different years is to plot them on so-called Kobe diagrams.[1] Figure 4.1 presents a Kobe diagram of the estimated positions of the global fisheries in 2004 and 2012, in terms of fishing effort and biomass. The vertical axis measures the current fishing effort relative to the optimal fishing effort that could be achieved in the long run—if this ratio is greater than unity, fishing effort exceeds what it could optimally be in the long run. The horizontal axis measures the current biomass relative to the optimal level that could be reached in the long run—if the ratio is less than unity, biomass must increase to reach its optimal value in the long run. When both ratios are equal to one, the fishery is at the long-run optimum, and the coordinate (1, 1) thus represents the long-run optimum of the fishery in a Kobe diagram.

The long-run optimal equilibrium of the fisheries is not the same for the two sets of conditions in 2004 and 2012, so the comparison is conducted in relation to the respective optimal points of the fisheries in the two different years. The

FIGURE 4.1

Kobe diagram for fishing effort and biomass, 2004 and 2012

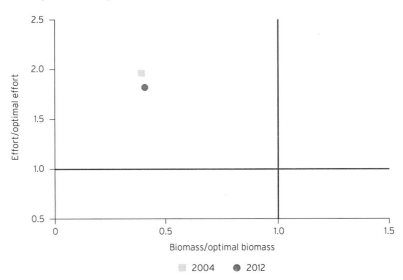

Source: Model output.

position of global fisheries in 2004 is assessed in relation to what would have been the fisheries optimum in 2004, while that same position for 2012 is compared to what fisheries optimum would have been for 2012.

As shown in figure 4.1, there has been an improvement in both effort and biomass since both have moved toward the optimal point (1, 1). When comparing the distance gained versus that which remains to reach the optimal point, however, it becomes apparent that it would take decades before global fisheries start approaching the optimal point.

Figure 4.2 presents a similar Kobe diagram that was drawn to measure the progress achieved in terms of net benefits and biomass, with similar results. Both benefits and biomass move in the direction of the long-run optimum, but at the current observed rate of progress, it would also take decades to reach the neighborhood of the optimal point.

It would appear from these two Kobe diagrams that the state of global fisheries has indeed improved over these eight years, but this slight improvement is not statistically significant at either the 5 percent or even 10 percent significance level. At the very least, then, the apparent positive evolution of global fisheries between 2004 and 2012 must be met with cautious optimism.

Different pathways to the optimal state for global fisheries

The model shows how the severely overexploited fish stocks have to be rebuilt over time according to the biological reproductive processes (which can some-

FIGURE 4.2
Kobe diagram for net benefits and biomass, 2004 and 2012

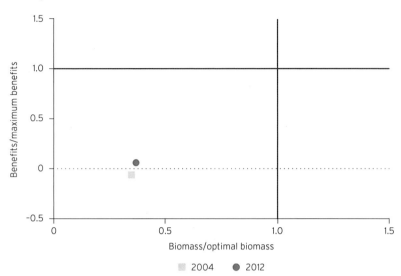

Source: Model output.

times be very slow). It also shows that to achieve fish-stock recovery, fishing mortality must be reduced through a reduction in fishing effort. According to the model, the global fishing effort would have to be reduced by 44 percent from that of 2012 (table 3.1) to reach the sustainable optimal state.

This chapter looks at various hypothetical pathways to achieve that state (at different rates), and asks what types of reforms would be required.[2]

According to standard economic investment theory, the optimal adjustment path is the one that maximizes the present value of the stream of benefits from the asset (Clark and Munro 1975; Solow 1956; Tobin 1969). The optimal path suggested by this bio-economic model for global fisheries is, theoretically at least, a most rapid path, or "bang-bang" solution (Clark and Munro 1975). However, there are very important social and political costs associated with altering the way fisheries are operated, particularly as the level of effort and number of fishers is modified. In absolute terms, an effective but radical strategy to improve the health of overexploited fish stocks would initially involve a temporary fishing moratorium, followed by the maintenance of an extended period of lower levels of fishing effort to achieve the needed reduction in harvest levels. This process would imply a temporary loss of income for all participants during a moratorium and the exit of some existing fishers from the sector thereafter. This process would, of course, create social and political tensions over a very large scale, which could be at least partly alleviated as part of a broader economic reform/development agenda. Any transition would have to be taken into consideration when designing fisheries transition policies. It should be noted, however, that in a number of countries a large share of fishers are among the poorest, living precariously in difficult circumstances and vulnerable areas, and with scant access to alternative livelihoods. These circumstances, in turn, may weigh heavily in the trade-offs that will decide the path of policy reform in each of those countries.

In addition to these important socioeconomic considerations, meaningful reform for the fisheries sector also raises physical and other practical difficulties. All fisheries operate with the underpinning of large amounts of capital, both physical capital in the form of vessels and fishing gear, and human capital in the form of experienced and trained labor. Much of this capital is imperfectly malleable, and as shown by Clark and Munro (1975), nonmalleable capital often signifies optimal adjustment paths that are no longer the most rapid ones but are more moderate ones, where fishing effort is allowed to evolve more smoothly over time toward its long-term optimal level.

Two hypothetical pathways are used here to illustrate and contrast the potential outcomes of alternative transition paths: the current path and a moderate path. The current path is generated by simulating the evolution that would take place if the fishing effort is kept current at the 2012 level. The moderate path is designed so that, starting from the observed 2012 level, the global fishing effort is gradually reduced at the annual rate of 5 percent from 2013 onward, until the long-run optimal level is attained.[3] The point here is not to delve into the details of transition policies for individual fisheries, but rather to illustrate the relationship between the speed of the transition and its benefits.

FIGURE 4.3
Assumed evolution of fishing effort under current and moderate paths

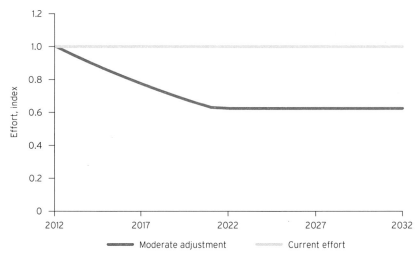

Source: Model assumption.

Figure 4.3 tracks the two paths for global levels of fishing effort. Fishing effort is measured as an index, whose 2012 value is set at unity, and the optimal fishing effort is reached by 2022 under the specified moderate adjustment path.

Figures 4.4 through 4.6 illustrate the *outcomes* of fisheries transition under the two assumed paths. According to these simulations, the moderate adjustment path does not lead to a substantial global harvest reduction in any given year (figure 4.4), but every year this level remains consistently below what it would be under the current policy. The fishing effort is reduced at a moderate rate under this path to allow the biomass to rebuild. The difference in harvest level between the two paths only begins to narrow by 2022, as the level of effort stabilizes and the stocks continue to recover quickly under the moderate path, while the harvest level reaches a plateau under the current path.

According to the model, a moderately fast reduction in fishing effort down to the optimal level would lead to a much higher global biomass level than if the global fishing effort were maintained at its 2012 level (figure 4.5). By 2032, the global biomass could reach 500 million tons, as opposed to around half that level under the current path.

However, it is when comparing fisheries net benefits that the difference between the two paths is the starkest (figure 4.6). Although it is indiscernible in figure 4.6, maintaining the 2012 fishing effort level would generate slightly higher net benefits during the first two years. But after that, the annual net benefits would increase quickly under the moderate path, even as fishing effort diminished. Thus, after the fishing effort reached its long-run dynamic optimum in 2022, and as the harvests begin to recover, global fisheries would

FIGURE 4.4
Evolution of harvest under current and moderate paths

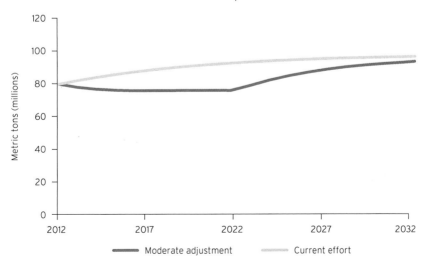

Source: Model output.

FIGURE 4.5
Evolution of biomass under current and moderate paths

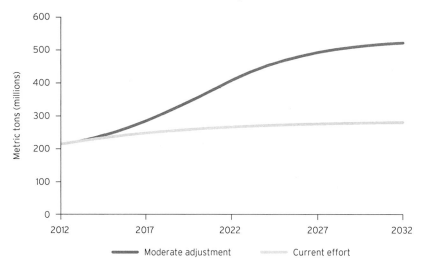

Source: Model output.

FIGURE 4.6
Evolution of fisheries net benefits under moderate and current paths

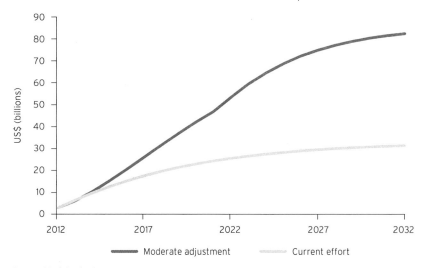

Source: Model output.

continue to reap the benefits of their transition. In contrast, the growth in benefits under the current path is markedly lower and the difference in net benefits between the two paths and the foregone benefits under the current path represent the strongest arguments in favor of reform carried out at a moderate pace.

TABLE 4.1
Estimated present value of global fishery, by adjustment
path (5 percent discount rate)

Adjustment path	Present value of fishery (US$ billions)
Most rapid path	1,465
Moderate path	1,196
Current path	514

Source: Model output.

Table 4.1 measures this difference in results between the two paths over 20 years. The present values of global marine fisheries are estimated by adding up the stream of discounted annual net benefits generated by the bio-economic model under the fishing effort specifications according to the adjustment paths analyzed in this subsection, and comparing it with the same results under the moderate path. A 5 percent discount rate is adopted. The table also shows the present value of the "bang-bang" approach, that is, reducing fishing effort to zero until fish stocks recover to the optimal level in about six years. As mentioned, this rapid path represents the near optimal path in terms of maximum present value of net benefits, but in reality it would be almost impossible to achieve.

According to these results, the maximum attainable present value of the global commercial marine fishing industry could theoretically reach close to $1.5 trillion, but at a considerable social and political cost. By implementing reform along a more moderate path, the attainable present *value* of net benefits would reach $1.2 trillion, or more than double the present value ($514 billion) of the current path. There is thus a compelling case for undertaking meaningful reform, centered mostly on a drastic, yet feasible, reduction in the level of effort.

A temporary reduction in benefits

Two major hurdles must be cleared in order to reach the optimal state of fisheries: first, a robust fisheries management system must be implemented; and second, the reduction to a more productive level of fishing effort requires that some fishers be transitioned out of the sector. Preparing and running such a fisheries management regime is costly and requires considerable technical and administrative capabilities that are not always available. Furthermore, the reduction in the level of fishing effort involves considerable social and economic adjustments, which are bound to create tensions, particularly among the fishers being displaced.

The first step toward overcoming these two hurdles is for fisheries managers and decision makers to understand the nature of the investment costs required to achieve the desired adjustment path. For most fisheries, it is possible to select an adjustment path that only reduces benefits by a relatively small amount at any given time compared to what would be achieved by continuing along

FIGURE 4.7
Assumed evolution of the fishing effort under the moderate and most rapid paths, in comparison with the current path

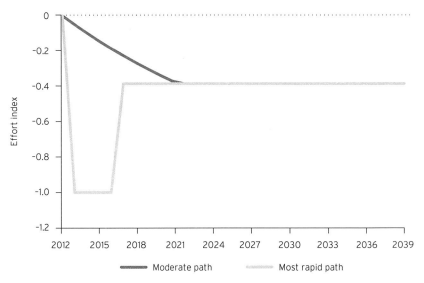

Source: Model assumption.

the current path. This approach's main drawback is that it will take comparatively more time before the increased net benefits begin to emerge, and a much longer time before the optimal sustainable state is attained, which in turn will decrease the discounted value of the stream of net benefits. The other extreme alternative is to select a very rapid adjustment path, which maximizes the present value of net benefits, but at very high short-term costs. The moderate adjustment path discussed above attempts to strike a balance between these two extremes. Figures 4.7 and 4.8 illustrate the various transition paths and their associated costs.

Figure 4.7 illustrates the two transition paths in comparison with the 2012 fishing effort level: the moderate path and the most rapid path, with the latter coming close to maximizing the present value of net benefits over time, but at a higher short-term cost. For both paths, fishing effort is depicted as deviating from the fishing effort assumed under the level that prevailed in 2012 (current path, traced at zero along the x axis). Under the most rapid path, fishing effort drops sharply for the first four years, followed by a net 40 percent reduction thereafter. Under the moderate path, in contrast, fishing effort diminishes more gradually, but eventually reaches the same long-term equilibrium level (close to 40 percent lower than fishing effort on the current path).

The most rapid transition path would result in a massive harvest shortfall during the first four years, when the long-run harvest would remain slightly

FIGURE 4.8
Evolution of net benefits under the moderate and most rapid paths, in comparison with
the current path

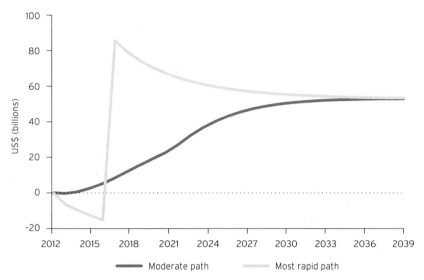

Source: Model output.

lower than under the current path. In contrast, under the more moderate path,
the initial shortfall in harvest would be much more modest, but would last over
15 years. In both cases, in the long term, the harvest level would remain slightly
below that of the current path (at the 2012 fishing effort level).

With regards to net benefits, the reduction in harvest also entails a steep
decline over the first four years of the adjustment period (figure 4.8) for the most
rapid path. The cumulative loss in net benefits over that period is estimated at
around $44 billion, but as the diagram shows, these initial losses are more than
compensated over the subsequent years. In fact, the very high net benefits in the
fifth year more than cover the cumulative losses of the first four. Nevertheless,
these initial losses are very real, and the initial $44 billion shortfall must be
covered, one way or another.

In contrast, and in the same figure, the moderate path avoids significant losses
in net benefits compared to the current path. Only during the adjustment period's first year does a comparatively small loss of $0.3 billion occur. Conversely,
however, it takes much longer for the neighborhood of optimal equilibrium to
be reached, and, as a result, the present value of the stream of net benefits under
the moderate path is about $270 billion lower than it would be under the most
rapid path (table 4.1).

In any event, and regardless of the adjustment path followed, a temporary
reduction in benefits will occur that must be compensated somehow.

Recovering the sunken billions: The way forward

Rationale

The model makes a very strong case for reform, but also clearly indicates that some costs will be incurred before the benefits can be derived. These results can be interpreted as follows: (a) that reform will require drastic reduction in the level of fishing effort, (b) which in turn will incur consequential costs, and (c) that the ultimate improvements in fisheries governance will generate sufficient net benefits to more than compensate these short-term costs. From an economic perspective, the issue is therefore twofold: first, these short-term costs must be borne (and represent a need for transitory funding), and second, and perhaps more importantly, the question arises of knowing who will be expected to bear these costs, particularly in contrast with who will ultimately reap the benefits of said reform.

The comparison between reform costs and benefits is made that much more arduous by the fact that all costs and benefits are not monetary in nature, and thus very difficult to compare. The benefits that can be derived from healthy, sustainably managed fisheries, and the ecosystems on which they depend, extend well beyond the price fetched from improved landings, including ecological services that are hard, if not impossible, to monetize. A growing body of research on the value of marine and coastal ecosystems is attempting, through economic valuation, to quantify the ways in which ecosystem services provide benefits to human populations, and expresses these values in monetary terms to allow for their comparison with other sources of societal value (Barbier and others 2009; UNEP-WCME 2011). This effort, however, is hindered by global climate change and the absence of reliable data (Costanza and others 2014), even as public awareness of these values continues to increase (Pendleton and others 2016). Nevertheless, awareness of the importance of these services is rising, and is making the case for protection and sustainable management of critical marine and coastal ecosystems even more clearly (World Bank 2016).

Beyond the exclusive focus on the monetary value of the reform costs and benefits, it is also helpful to look at fisheries as a mode of natural capital exploitation, including to assess whether or not this exploitation is sustainable. One recommendation of the original version of *Sunken Billions* was to value fish wealth as natural capital, and to attempt to measure that level of wealth in the national accounts of countries (World Bank and FAO 2009). A first step to that end was taken by the World Bank in Mauritania, where a wealth accounting analysis was conducted, which showed that natural capital accounts for 44 percent of the country's total stock of natural wealth, and fisheries account for just over one-quarter of the country's natural capital (see box 4.1 and Mele [2014]). Similarly, a model was developed with the Central Bank of Morocco as a basis for the creation of a national fisheries account for that country.

Public expenditures are required to cover the cost of fisheries management and enforcement. This cost can be considerable, but represents a necessary invest-

BOX 4.1
Accounting for fish wealth in Mauritania

In 2014, the World Bank conducted an analysis of Mauritania's existing capital stock to assess the country's produced, intangible, and natural capital—both renewable and nonrenewable. The study results showed that the country's stock of natural capital amounts to approximately US$30–35 billion, or roughly US$9,000 per capita, and represents 44 percent of the total capital. More than half of the country's natural wealth is concentrated in renewable resources, which, given effective sustainability-focused policies, could theoretically support a continuous income flow over the long term. Such sustainable management is not a given, however, and unsustainable management of renewable resources can lead to permanent depletion of capital stocks in much the same way as the finite extraction of nonrenewable resources.

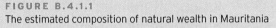

FIGURE B.4.1.1
The estimated composition of natural wealth in Mauritania

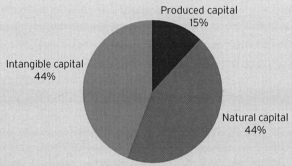

Produced capital
15%

Intangible capital
44%

Natural capital
44%

Source: Mele 2014.

   The fisheries sector represents the largest share of natural wealth in Mauritania, and with an estimated US$10 billion value (or roughly US$2,800 per capita), fisheries account for just over a quarter of the country's natural capital. Commercial fishing represents approximately 90 percent of the sector, with artisanal fishing accounting for the remaining 10 percent. Fishing contributes just 3 percent to annual GDP, but the sector registered double-digit growth rates in 2010 and 2011, and is expected to grow by 5 percent per year over the medium term. Meanwhile, the revenues generated by international fishing agreements have remained roughly constant as a share of total revenues for much of the past seven years, typically accounting for around 20 percent of public sector income.
   However, the presence of significant natural wealth does not always translate into shared prosperity, either for the current population or for future generations. The failure to responsibly manage natural resources and adopt policies that expand the economic impact of resource exploitation can jeopardize broad-based growth and poverty reduction both now and in the future. Some of Mauritania's local fish species are already overexploited, and widespread global overfishing is expected to boost the value of remaining fishery resources. This situation underscores the importance of optimizing rents from commercial fisheries and using a share of those resources to ensure that the sector is properly regulated. Without effective monitoring and enforcement, overfishing of the highest-value species (for example, octopus) may seriously jeopardize the regenerating mechanisms of the country's fisheries. If left unchecked, overexploitation could transform the fisheries sector into a so-called "finite resource," which is a renewable resource that is rendered nonrenewable and thus may become biologically or commercially extinct over time.

Source: Mele 2014.

ment in the sustainable exploitation of a common good, designed to ensure an unending flow of economic benefits and as a critical means to ensure food security. The cost then becomes part of the "institutional fabric of fisheries tenure at all levels" (World Bank and FAO 2009), and the reform cost is but another component of the cost of managing fisheries, where the trade-off, or allocation of these costs between the public and private sectors, must be decided from the perspectives of fiscal policy and equitable burden sharing.

The paths to reform

While it makes a compelling case for reform, and for recovering the sunken billions lost every year, this study does not take a prescriptive approach to this reform. There is no single approach to address the overcapacity that is at the root of the problem. Some key factors, such as the open-access regime that prevails overwhelmingly, or the negative impacts of perverse subsidies, appear to play an important role in many countries, but *how* these factors are addressed needs to be country-specific. Very different approaches have proved to be successful in addressing these and other related issues, yet they vary greatly, depending on the local circumstances. Even within individual countries, different fisheries management regimes sometimes prevail. Nevertheless, it is useful to consider and evaluate the different paths to reform along a series of key considerations.

Reevaluating subsidies

Globally, significant efforts are already under way to tackle the negative effects of distorting subsidies, including through the World Trade Organization process (Sumaila 2016). Perverse fishing subsidies, which distort the true costs of fishing and artificially inflate the rent that can be derived, are major contributors to overcapacity, making fishing more attractive economically than it really is and ultimately fueling overfishing. However, global subsidies are significant: some recent studies estimate global fisheries subsidies at US$35 billion a year, around US$20 billion of which are provided in forms that tend to further increase fishing capacity (Sumaila and others 2010; Sumaila and others 2012; and Sumaila and others 2016). This subsidy level is equivalent to as much as one-third of the value of global fisheries production. The negative impact of fishing over-capacity and subsidies on current fish stocks—and therefore on fisheries in the medial and long terms—are now widely recognized worldwide and are embedded in the United Nations' Sustainable Development Goals (Goal 14.6).[4]

Any effort toward reducing the scope of these subsidies will have two immediate effects, namely: achieving fiscal savings due to reduced transfers to the fishing sector, and redressing the economic distortions they create, thus reducing the drive toward overcapacity and overfishing. This situation does not necessarily mean that the governments engaging in such subsidization need to reduce their sector investment and their support to fishers who are most often the intended beneficiaries. Rather, a very positive first step could be for them to redirect their support along a more productive path. For instance, these funds could be diverted toward supporting the fishers expected initially to reduce their level of fishing and even in some cases to exit the fisheries sector altogether.

Technical and regulatory approaches

A major 2009 study on rebuilding global fisheries systematically reviewed the kind of approaches that can be taken and how their results differed (Worm and others 2009). The study clearly showed that all approaches were not appropriate everywhere and that the same kind of approach could have different results depending on where they were implemented. The study reviewed these interventions around eight broad categories, as follows:

- Gear restrictions, which can be used to increase selectivity and reduce by-catch of nontarget species
- Closed areas—fully protected or designed around core areas outside which some restricted uses are allowed—allow species to recover, restore community structure, protect critical habitats, and increase ecosystem resilience
- Reduction in total allowable catch, set closer to maximum sustainable yield (MSY) or maximum economic yield (MEY) levels
- Reduction in total fishing effort, for instance around closed seasons
- Community co-management, or community-based resource management
- Reduction in capacity, through a reduction in the number of licenses or buy-back schemes
- Catch shares, rights-based management, or territorial fishing rights, where dedicated access privileges are assigned to individual fishers or fishing communities, with a view to providing economic incentives to reduce effort and exploitation rate
- Often in combination with the other measures above, fisheries certification schemes, where improved management practices are rewarded with better access to higher-value markets.

The effects of these measures, implemented individually or in combination, were then evaluated over different ecosystems and in 10 very different regions, and varied greatly, depending on the fisheries, ecosystem, and governance system. Catch shares were recognized as being particularly efficient in some circumstances, while in others, community-based management was a better predictor of success. The study also showed that rather than any one single approach, a combination of diverse tools was often necessary to rebuild the targeted fisheries.

Notwithstanding these differences and uncertainties, successful approaches irrefutably showed that overfishing can be addressed and fisheries rebuilt.

Leveling the playing field—tackling illegal, unregulated, and unreported fishing

At the global level, however, individual countries are not always able to impose the necessary conservation and management measures, particularly in the case of foreign fishing agreements (FFAs), where distant water fishing nations are able to negotiate favorable terms and the coastal states are not always able to enforce the measures already in place (Alder and Sumaila 2004). According to a 2014 World Bank study, approximately half of the world's Exclusive Economic Zones (EEZs) are subject to some form of FFA, and some countries fared much better in imposing strict management and control measures, particularly when they negotiated as a block as Pacific Island countries have been able to do (World Bank 2014).

This inability to impose the necessary conservation and management measures is at the heart of the IUU fishing issue (illegal, unregulated, and unreported fishing, discussed in greater detail in chapter 1). None of the possible approaches can be effective unless they are implemented and complied with, and, in case of noncompliance, sufficient enforcement measures are in place. By definition, distant water fishing fleets are mobile and, when intent on avoiding these stronger measures, will naturally migrate to other areas where the laws are weaker or enforcement is lacking. This situation is particularly the case of areas beyond national jurisdiction, on the high seas and outside the EEZs of coastal nations, where, in the absence of the exclusive rights granted coastal states in their EEZs under the Law of the Sea Convention, conservation and management measures are negotiated and adopted by regional fisheries management organizations, with varying degrees of success (Cullis-Suzuki and Pauly 2010; de Fontaubert and others 2003).

Preparing for climate change impacts

Finally, global fisheries reform designed to recover the sunken billions will also make fisheries, related coastal and marine ecosystems, and, ultimately, fishing communities more resilient to exogenous shocks, including those associated with climate change. As the effects of sea level rise, ocean acidification, changes in currents, and other changes in ocean systems increase, their inevitable impacts will become more pronounced, making adaptation that much more difficult (Alison and others 2009; Barbier 2015). Studies have clearly demonstrated the economic cost that can be expected on both coastal areas (Nicholls and others 2007), and fisheries (Daw and others 2009). And, as noted in chapter 1, the advent of these impacts removes the natural buffer under which overfished stocks stand a biological chance to recover, if and when their overexploitation ceases, as a result of the projected biomass increases. In turn, rebuilding fisheries natural capital, in the form of biologically resilient fish stocks and coastal ecosystems that provide key habitats to marine life and natural protection to coastal communities, makes coastal communities that much more resilient to physical shocks. As climate change impacts on oceans become more severe, the need for comprehensive fisheries reform at a global scale becomes more pressing than ever.

Notes

1. The diagrams are named after Kobe, Japan, where a meeting of the five major tuna fisheries management organizations first designed and adopted them as a presentation method.

2. Some of the most innovative and comprehensive work on rebuilding fisheries was carried out by the OECD (2012).

3. This long-run optimal level is a *dynamic* optimum, where the discounted sum of the stream of annual net benefits is maximized at an annual 5 percent discount rate. Note that this long-run optimal level is close, but not identical, to the *static* optimal equilibrium at maximum economic yield discussed in the previous chapters (for example, table 3.1). Specifically, the dynamic optimum involves a slightly higher fishing effort and harvest and lower biomass in the long run than the static optimal equilibrium, where the sustainable net benefits from the fishery are maximized (see, for example, Clark and Munro [1975]).

4. Sustainable Development Goal 14.6 reads: "By 2020, prohibit certain forms of fisheries subsidies which contribute to overcapacity and overfishing, eliminate subsidies that contribute to illegal, unreported and unregulated fishing and refrain from introducing new such subsidies, recognizing that appropriate and effective special and differential treatment for developing and least developed countries should be an integral part of the World Trade Organization fisheries subsidies negotiation."

References

Alder, J., and U. R. Sumaila. 2004. "Western Africa: A Fish Basket of Europe Past and Present." *Journal of Environment and Development* 13 (2): 156–78.

Alison, E. H., A. L. Perry, M. C. Badjeck, W. N. Adger, K. Brown, D. Conway, A. S. Halls, G. M. Pilling, J. D. Reynolds, N. L. Andrew, and N. K. Dulvy. 2009. "Vulnerability of National Economies to the Impacts of Climate Change on Fisheries." *Fish and Fisheries*. http://www.uba.ar/cambioclimatico/download/Allison%20et%20al%202009.pdf.

Barbier, E. B. 2015. "Climate Change Impacts on Rural Poverty in Low-Elevation Coastal Zones." *Estuarine, Coastal and Shelf Science* 165 (June): 1–13.

Barbier, E. B., and others. 2008. "Coastal Ecosystem-Based Management with Nonlinear Ecological Functions and Values." *Science* 319 (5861): 321–23.

Barbier, E. B., S. D. Hacker, C. Kennedy, E. W. Koch, A. C. Stier, and B. R. Silliman. 2009. "The Value of Estuarine and Coastal Ecosystem Services." *Ecological Monographs* 81: 169–93. doi:10.1890/10-1510.1.

Clark, C., and G. R. Munro. 1975. "The Economics of Fishing and Modern Capital Theory: A Simplified Approach." *Journal of Environmental Economics and Management* 2: 92–106.

Costanza, R., R. de Groot, P. Sutton, S. van der Ploeg, S. J. Anderson, I. Kubiszewski, S. Farber, R. K. Turner. 2014. "Changes in the Global Value of Ecosystem Services." *Global Environmental Change.* doi:10.1016/j.gloenvcha.2014.04.002.

Cullis-Suzuki, S., and D. Pauly. 2010. "Failing the High Seas: A Global Evaluation of Regional Fisheries Management Organizations." *Marine Policy* 34 (5): 1036–42.

Daw, T., W. N. Adger, K. Brown, and M-C. Badjeck. 2009. "Climate Change and Capture Fisheries: Potential Impacts, Adaptation and Mitigation." In "Climate Change Implications for Fisheries and Aquaculture: Overview of Current Scientific Knowledge," edited by K. Cochrane, C. De Young, D. Soto, and T. Bahri, 107–50. FAO Fisheries and Aquaculture Technical Paper No. 530, FAO, Rome.

de Fontaubert, C. and I. Lutchman, with D. Downes and C. Deere. 2003. "Achieving Sustainable Fisheries: Implementing the New International Legal Regime." IUCN, Gland, Switzerland, and Cambridge, U.K.

Mele, Gianluca. 2014. "Mauritania Economic Update: 1." World Bank, Washington, DC. http://documents.worldbank.org/curated/en/2014/07/20133610/mauritania-economic-update.

Nicholls, R. J., P. P. Wong, V. R. Burkett, J. O. Codignotto, J. E. Hay, R. F. McLean, S. Ragoonaden, and C. D. Woodroffe. 2007. "Coastal Systems and Low-Lying Areas. Climate Change 2007: Impacts, Adaptation and Vulnerability. Contribution of Working Group II to the Fourth Assessment Report of the Intergovernmental Panel on Climate Change," edited by M. L. Parry, O. F. Canziani, J. P. Palutikof, P. J. van der Linden, and C. E. Hanson. Cambridge University Press, Cambridge, U.K., 315–56.

OECD (Organisation for Economic Co-operation and Development). 2012. *Rebuilding Fisheries: The Way Forward.* Paris: OECD Publishing. http://dx.doi.org/10.1787/9789264176935-en.

Pendleton, L. H., O. Thebaud, R. C. Mongruel, and H. Levrel. 2016. "Has the Value of Global Marine and Coastal Ecosystems Services Changed?" *Marine Policy* 64: 156–58.

Solow, R. M. 1956. "A Contribution to the Theory of Economic Growth." *Quarterly Journal of Economics* 70(1): 65–94.

Sumaila, R., W. Cheung, A. Dyck, K. Gueye, L. Huang, V. Lam, D. Pauly, T. Srinivasan, W. Swartz, and R. Watson. 2012. "Benefits of Rebuilding Global Marine Fisheries Outweigh Costs." *PLoS ONE* 7 (7): e40542.

Sumaila, R., A. Khan, A. Dyck, R. Watson, G. Munro, P. Tyedmers, and D. Pauly. 2010. "A Bottom-Up Re-Estimation of Global Fisheries Subsidies." *Journal of Bioeconomics* 12: 201–25.

Sumaila, R., and U. Rashid. 2016. "Trade Policy Options for Sustainable Oceans and Fisheries. E15 Expert Group on Oceans, Fisheries and the Trade System – Policy Options Paper" E15 Initiative, International Centre for Trade and Sustainable Development (ICTSD) and World Economic Forum, Geneva.

Tobin, J. 1969. "A General Equilibrium Approach to Monetary Theory." *Journal of Money, Credit and Banking* 1 (1): 15–29.

UNEP-WCMC (United Nations Environment Programme–World Conservation Monitoring Centre). 2011. "Marine and coastal ecosystem services: Valuation methods and their application." UNEP-WCMC Biodiversity Series No. 33. UNEP-WCMC, Cambridge, UK.

World Bank. 2014. "Trade in Fishing Services: Emerging Perspectives on Foreign Fishing Arrangements." World Bank, Washington, DC.

———. 2016. "Managing Coasts with Natural Solutions: Guidelines for Measuring and Valuing the Coastal Protection Services of Mangroves and Coral Reefs," edited by M. W. Beck and

G-M Lange. Wealth Accounting and the Valuation of Ecosystem Services Partnership (WAVES), World Bank, Washington, DC.

World Bank and FAO (Food and Agriculture Organization). 2009. *The Sunken Billions: The Economic Justification for Fisheries Reform.* Washington, DC: World Bank.

Worm, B., R. Hilborn, J. Baum, T. Branch, and others. 2009. "Rebuilding Global Fisheries." *Science* 325 (5940): 578–85.

Basic Approach and Methodology

The fundamental approach of this work consists of a few steps, as follows:

1. Global marine fisheries are treated as one large fishery.

2. This fishery is described by a bio-economic model that is consistent with fisheries economics theory and empirical knowledge about the global fishery.

3. The bio-economic model is made operational by empirical estimates of its parameters producing the estimated bio-economic model.

4. Empirical information about the global fishery state in the base year is combined with the estimated bio-economic model to obtain estimates of the base-year net benefits.

5. The estimated bio-economic model is used to calculate the maximum sustainable net benefits attainable from the global fishery.

6. The difference between the maximum net benefits attainable on a sustainable basis and the current benefits represents the foregone economic benefits, here referred to as the "sunken billions." As clarified in footnote 1 of chapter 1, due to data constraints, these net benefits are neither strictly net financial benefits (profits) nor net economic benefits. Rather, in the modeling undertaken in this report, net benefits represent an in-between approximation.

7. Stochastic simulation methods are employed to obtain confidence intervals for the sunken billions estimate.

8. The estimated bio-economic model is used to evaluate recovery paths from the base-year state of the fishery to the long-run economic benefits maximizing level.

The bio-economic model

With a few modifications, this study employs a bio-economic model that is similar to the one that was used in the original *Sunken Billions* study (Arnason 2011; World Bank and FAO 2009). The model is a typical aggregative fisheries model with specifications in accordance with accepted fisheries economics theory and empirical knowledge (see for example, Anderson [1977], Anderson and Seijo [2010], Bjorndal and Munro [2012)], and Clark [1980]). Appendixes B and C describe in some detail the model and the way its parameters are estimated. This appendix reviews the model's main features.

Naturally, the model represents a highly simplified description of the global fishery. Most crucially, the model assumes that global fisheries can be modeled as a single fish stock with one aggregate biomass growth function. Similarly, the global fishing industry is represented by an aggregate fisheries profit function, composed of an aggregate harvesting function, relating the harvest to fishing effort and biomass, and an aggregate cost function relating fishing effort to fisheries costs. Finally, the landed catch price is represented by a single price function that depends positively on the global fish stock biomass.

Thus, this bio-economic model contains four fundamental functional relationships, as follows: (i) a biomass growth function describing the natural growth of the fish stock biomass as a function of the biomass itself; (ii) a harvesting function describing the landed catch volume as a function of the biomass and fishing effort; (iii) a fishery cost function describing the fishing cost as a function of fishing effort; and, (iv) a fish price function describing the landed catch price as a function of biomass. All these functions include various parameters that need to be estimated (see subsequent discussion and appendix C).

The model is dynamic, in that it can describe the global fishery evolution over time. Fundamental to this evolution is the biomass that grows or declines with the difference between natural biomass growth and harvest. However, the other major source of dynamics in fisheries, fishing capital, is not included in this model. The model dynamics are of a discrete nature. There is biomass at the beginning of the year, which is increased by biomass growth and reduced by harvest, leading to a new biomass at the beginning of the next year. If that biomass is the same as it was the year before, an equilibrium, or sustainable state, is reached, and otherwise the fishery evolves further.

The model is also stochastic. The stochasticity enters the model via probability distributions specified for the model parameters. Thus, model outcomes, including net benefits, also have probability distributions and the point estimates are subject to confidence intervals.

The model contains eight basic variables, as follows: (i) biomass; (ii) biomass growth; (iii) harvest volume; (iv) price of landed catch; (v) revenues; (vi) fishing costs; (vii) net benefits; and (viii) fishing effort. The first seven of these variables are determined endogenously by the model, given the values of the parameters and the initial biomass. The eighth variable, fishing effort, is the model's exogenous driver. Humans can control fishing effort, and the fishing effort selected determines the net benefits obtained from the fishery as well as its evolution over time. In the global fishery as it is currently organized, certain socioeconomic processes determine fishing effort. This fishing effort generates certain biological and economic outcomes, which previous estimates indicate are severely unfavorable. Alternatively, fishing effort may also be set so as to maximize sustainable net benefits. Comparing the two outcomes provides an estimate of the sunken billions, which are the net benefits that could have been accrued, but instead are wasted in the global marine fishery.

It may be helpful at this stage to describe the basic operation of the bioeconomic model through a flowchart that explains how the model's endogenous variables are generated, including net benefits, as shown in figure A.1.

At the beginning of each year, there is a certain biomass. At this time some fishing effort is selected, which leads to fishing costs according to the cost function

FIGURE A.1
Bio-economic model flowchart

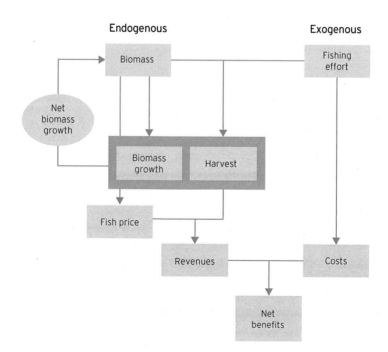

as illustrated in the chart. Further, fishing effort, in combination with biomass, produces harvest, as specified by the harvesting function. Biomass also generates a landed catch price according to the price function. Multiplying the harvest with the landed catch price generates revenues. The difference between the revenues and fishing costs produces the net benefits from the fishery during this period.

As well as leading to net benefits during the year, the fishing effort that is selected produces, in combination with other variables, the biomass at the beginning of the next period and, thus, the evolution of the fishery over time. As illustrated in figure A.1, subtracting the harvest from the natural biomass growth generates net biomass growth. Adding this net biomass growth—which may be positive or negative—to the initial biomass produces the biomass at the beginning of the next year. If that biomass is the same as the initial biomass, the fishery is in a sustainable state. If not, biomass, and therefore net benefits as well, continue to evolve.

In the model's stochastic version, the described process does not produce deterministic outcomes from any given fishing effort selected. Instead the outcomes are stochastic, or uncertain, and have to be described by a probability distribution.

A common way to provide further insight into the nature of bio-economic models is to draw a diagram of their steady state or sustainable outcomes. Figure A.2 represents the study's bio-economic model. Note that figure A.2 is drawn for specific values of the parameters. Other parameter values will produce a different diagram although the key features of figure A.2 will persist.

FIGURE A.2
The bio-economic model: A steady state representation

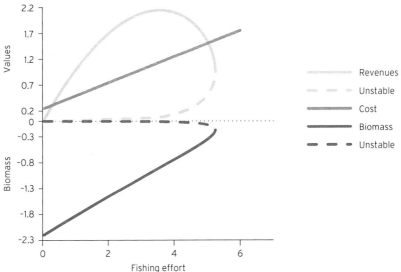

Note: Biomass is measured in a downward direction.

Figure A.2 provides an example of the sustainable biomass and economic outcomes from the fishery as functions of fishing effort resulting from the study's bio-economic model. The economic outcomes, sustainable revenues, and costs are depicted in the upper half of the diagram and the sustainable biomass in the lower half. Note that the biomass volume is measured in a downward direction, so the lower on the vertical axis, the higher the biomass.

The fishing cost function is a simple one with certain fixed costs (the intercept on the vertical axis) and variable costs increasing proportionately with fishing effort. The sustainable revenue and biomass functions are more complicated. Sustainable revenues initially increase with fishing effort as harvests are increased. However, since increased fishing effort reduces biomass, the rate of revenue increase gradually peters out until at some point of fishing effort (approximately 3.7 in figure A.2), the maximum sustainable revenues are reached. Note that since revenues are a multiple of the price and volume of landings, the fishing effort corresponding to the maximum sustainable revenues will not, in general, coincide with the fishing effort corresponding to the maximum sustainable yield (MSY). For any fishing effort beyond the one corresponding to maximum sustainable revenues, revenues are reduced. At a certain fishing effort level (approximately 5.3 in figure A.2), which may be referred to as the critical effort, the sustainable revenue curve exhibits a discontinuity. In technical terms it suffers a bifurcation (Clark 1990), which means that the nature of the relationship is transformed. A sustainable revenue curve still exists, indicated by the dashed curve in the diagram, but the revenues are much lower and the points on the curve are dynamically unstable. The point of discontinuity (or bifurcation) has the practical implication that if fishing effort is maintained above the critical level, sustainable revenues and harvest drop from a significantly positive level (as in figure A.2) to zero. Thus the critical effort level may be seen as being at the very edge of the fisheries precipice.

The sustainable biomass curve exhibits the same critical features as the sustainable revenue curve. It falls monotonically with fishing effort until the critical effort level is reached. At that point, the sustainable biomass curve exhibits a discontinuity (or bifurcation) and is replaced by an unstable arm (indicated by the dashed curve in the diagram). The discontinuity in the sustainable biomass at the critical effort occurs at a significantly positive biomass level and, thus, illustrates the same collapse in sustainability as in the case of the revenue curve.

In figure A.2, it is worth drawing attention to the asymmetry of the sustainable revenue curve. Often, in sustainable fishery diagrams the revenue (or yield curve) are drawn as symmetric curves—obviously a special case that is unlikely to happen in reality. The study's bio-economic model is flexible enough to offer the possibility of a nonsymmetric sustainable revenue curve, depending on the parameter values selected. Several model components create this flexibility. The biomass growth function selected—the Pella-Tomlinson function (appendix B; Pella and Tomlinson 1969)—plays a role in this, as do the fish harvesting function with its schooling parameter and the fish landing price function (see appendix B).

FIGURE A.3

The bio-economic model: A dynamic representation

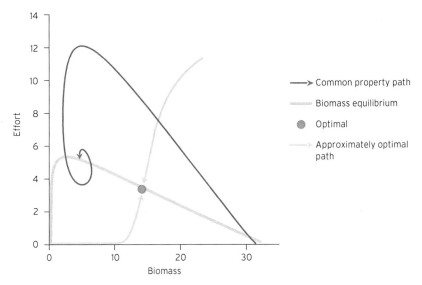

Source: Model.

Another way to gain insight into the workings of the bio-model is to study its dynamic properties, which can be done with the help of the so-called state-space representation of the model. State-space representation is a technical term for a particular way to depict a model's dynamic properties. Figure A.3 presents a state-space illustration of the study's bio-economic model in the space of effort (vertical axis) and biomass (horizontal axis). The solid bold curve represents combinations of fishing effort and biomass that maintain a constant biomass. In other words, it is the sustainable biomass curve of figure A.2 with the axes reversed. The highest point of this curve corresponds to the critical effort level in the sustainable biomass curve in figure A.2. For certain parameter values this maximum exists, for others it doesn't. Whenever fishing effort is below the sustainable biomass curve, the biomass is increasing and vice versa. This figure indicates the evolution of biomass in this effort-biomass space. The evolution of fishing effort, on the other hand, depends on the fisheries management system in place.

Figure A.2 considers two possible evolutions of fishing effort. One represents the common property or common pool situation, under which fishing effort expands whenever net benefits are positive (that is, profits), and contracts when they are negative. The other represents the optimal effort management, which sets fishing effort at each point in time so as to maximize the present value of net benefits from the fishery. These two policies give rise to the fishery's widely different evolutionary paths. The first path (labeled "common property path") describes the evolution of the common property fishery from a quite underutilized state (biomass approximately 31 units) and low initial fishing effort (approximately 1 unit) toward the long-run equilibrium of no profits. As illustrated, this cyclical evolutionary path has fishing effort increasing when profits are positive and declining when they are negative. This type of evolutionary path for a common property fishery appears to fit with observed reality (Anderson and Seijo 2010; Bjorndal 1987; Wilen 1976). Note that in figure A.2, the equilibrium point to which the fishery converges corresponds to the zero profit equilibrium in the static representation. This point is characterized by low biomass, high fishing effort, and zero net benefits from the fishery, and is fairly close to the critical effort level (point of bifurcation) and therefore quite risky.

The other evolutionary curve (labeled "approximately optimal path") specifies the path of fishing effort that would approximately maximize the fishery's present value of net benefits. This path converges to the long-run optimal equilibrium (labeled "optimal" in figure A.3). Importantly, this optimal long-run equilibrium is generally not identical with the optimal static point. Usually it corresponds to a slightly lower biomass (Clark and Munro 1975). As indicated in figure A.3, when biomass is low compared to the long-run optimum, it would be optimal to use very little or no fishing effort to increase the biomass as fast as possible. When biomass is relatively high compared to the long-run optimum, fishing effort should be relatively high to take the best advantage of high biomass to generate net benefits. The fishery's optimal dynamics are thus totally different from the dynamics under the common property arrangement. The fishery's long-run equilibrium or sustainable states are also widely different. The equilibrium under the common property arrangement, as already mentioned, is characterized by high fishing effort and low biomass. The optimal sustainable state, by contrast, is characterized by comparatively high biomass and low fishing effort.

Figure A.3 illustrates one possible set of fishery evolutions over time toward (i) common property and (ii) optimal long-run equilibria. While any fishery's actual evolution depends on the value of the model parameters and the state of the fishery in the first year, it will be qualitatively similar to the ones depicted in figure A.3. The crucial message of the dynamics depicted in figure A.3 is that the evolutionary paths take time, irrespective of whether they are toward the long-run optimum or to the common property point of no net benefits. In particular, it is not possible to jump immediately to the long-run optimum. Indeed, it may take many years to rebuild fish stocks to the optimal equilibrium, especially if the initial overexploitation is great. During this time of stock rebuilding the fishery may suffer losses or generate net benefits, which clearly must be taken into account in designing the optimal global fisheries policy.

Net benefits

In theory, the net benefits derived from fishing are defined as the social value of landed catch minus the social value of the inputs used to produce the catch. In economic theory, the social value of anything, including inputs and outputs, equals the quantity in question multiplied by the "true" price (Debreu 1959).

The true price is the one that accurately measures the benefits (or costs) of one more unit of the quantity. In a perfect market system all prices are true in this sense (Arrow and Hahn 1971; Debreu 1959), but in real market systems, actual prices will generally not coincide with true prices, although the difference may not necessarily be great.

Fishing's net benefits are estimated as the difference between fishing revenues and fishing costs evaluated at the 2012 base-year prices, netting out taxes and revenues (which are transfer payments and not true costs of production). For this to be a measure of net benefits, 2012 market prices must be true in the above sense, whereas this is not the case in this report due to poor data on both the side of production costs and government subsidies/taxes. As a result, the difference between fisheries revenues and fishing costs estimated here is not a precise but rather an approximate measure of net benefits that falls between a more strict financial and economic valuation.

The price of landings is assumed to be independent of their volume. This is a simplifying assumption adopted to avoid certain technical difficulties in locating the benefit maximizing fishery.[1]

Maximizing net benefits

The crucial features of the bio-economic model in chapter 2 are that the global fishing effort is exogenous and can be selected, while the rest of the model endogenously generates the consequences of that selection for the global fishery. Thus, at the beginning of each year, and once a level of fishing effort has been selected, the model generates a set of fishery outcomes including the volume of harvest, net benefits, and, most importantly, the biomass remaining at the beginning of the next period.

The inherent difficulty is to select a level of fishing effort so as to maximize the net benefits derived from the fishery. If fishing effort is unwisely selected, the net benefits will be low, or even negative, and the fish biomass reduced. If fishing effort is wisely selected, the fish stocks will be healthier, and high net benefits will be derived from the global fishery on a sustainable basis.

Identifying the fishing effort that maximizes the fishery's net benefits is, per se, a purely technical problem, but adopting and implementing a solution is primarily a social and political problem. This section is only concerned with the technical aspect of the problem; the social and political aspects are not considered here.

With the estimated bio-economic model in hand, it is fairly straightforward to locate the fishing effort that maximizes the sustainable net benefits. Because of the nonlinear nature of bio-economic fisheries models, the maximizing solution cannot be expressed as an equation (Clark 1990). Instead, to locate the solution, a numerical search has to be conducted. There are many methods to conduct this search, and in this particular case, the search was conducted over the space of sustainable biomasses until the one corresponding to the highest possible net benefits was reached. Since the point that maximizes sustainable net benefits from the fishery does not really take the passage of time into account, it may be referred to as the static optimal equilibrium.

Locating the path of fishing effort over time that maximizes the present value of net benefits is an exercise in dynamic maximization, which is inherently complicated and was applied in two steps. First, findings from dynamic optimization of general fisheries models were used to narrow down the set of possible optimal fishing effort paths. Second, a numerical search was conducted over the set of possible paths of fishing effort to locate one that approximately solved the problem.

It was found that the path of fishing effort that maximizes the present value of net benefits from the global fishery converges to a sustainable fishery, that is, an equilibrium where biomass stays constant, which may be referred to as the dynamic optimal equilibrium. It is important to note that the dynamic optimal equilibrium is generally not at the same point as the static optimal equilibrium, that is, the sustainable fishery which maximizes the flow of net benefits. Generally, the dynamic optimal equilibrium implies slightly higher fishing effort and lower biomass than the static optimal one. The normally small difference is caused by the positive discount rate used in the present value calculations.

Accounting for model inaccuracy

The above information makes it clear that the approach to the global fishery modeling suffers from many inaccuracies, which include the following: the model itself is a very imperfect description of the global fishery; the model parameters are estimated on the basis of limited data, bringing in additional inaccuracies; and, the estimation of net benefits as the difference between fishery revenues and costs is at best an approximation. As a result, the key outcomes of this study, including the estimate of the foregone economic benefits, that is, the sunken billions, in the global marine fishery in 2012, are subject to considerable inaccuracy.

In an attempt to account for this inaccuracy, two methods were used: First, an analysis was conducted of the sensitivity of the key results to several of the modeling assumptions, which provided an idea of the possible range of actual 2012 foregone benefits in the global fishery. Second, stochastic simulation methods were employed to generate confidence intervals for the actual foregone benefits in the global fishery. This calculation was done by specifying probability distributions for various model parameter estimates and assumptions, and then repeatedly drawing values from these distributions to generate a probability distribution for the foregone benefits. This method, generally referred as Monte Carlo simulations, can generate probability distributions for the outcomes of any precision that is desired. The empirical reliability of these distributions, however, is only as good as the appropriateness of the stochastic specifications of the model parameters and other components.

Despite their usefulness, both of these approaches are limited and cannot fully account for the uncertainty about the true value of the foregone economic benefits of the global fishery in 2012. They do, however, provide an added insight into the likely range of these benefits.

Note

1. If the landed catch price depends on the volume of landings, simple maximization of net benefits will make the fishery act as a global monopoly ignoring consumer benefits (Varian 1992).

References

Anderson, L. 1977. "The Economics of Fisheries Management." Johns Hopkins University Press, Baltimore.

Anderson, L. and J. Seijo. 2010. *Bioeconomics of Fisheries Management.* New York: John Wiley & Sons.

Arnason, R. 2011. "Loss of Economic Rents in the Global Fishery." *Journal of Bioeconomics* 13: 213–32.

Arrow, K., and F. Hahn. 1971. *General Competitive Analysis.* North Holland.

Bjorndal, T. 1987. "Production Economics and Optimal Stock Size in a North Atlantic Fishery." *Scandinavian Journal of Economics* 89: 145–64.

Bjorndal, T., and G. Munro. 2012. *The Economics and Management of World Fisheries.* U.S.: Oxford University Press.

Clark, C. W. 1980. "Towards a Predictive Model for the Regulation of Fisheries." *Canadian Journal of Fisheries and Aquatic Science* 37: 1111–129.

———. 1990. *Mathematical Bioeconomics: The Optimal Management of Renewable Resources.* 2nd ed. New York: John Wiley & Sons.

Clark, C., and G. Munro. 1975. "The Economics of Fishing and Modern Capital Theory: A Simplified Approach." *Journal of Environmental Economics and Management* 2: 92–106.

Debreu, G. 1959. *Theory of Value.* Cowles Foundation, Monograph 17. New Haven: Yale University Press.

Pella, J. J., and P. K. Tomlinson. 1969. "A Generalized Stock Production Model." *Inter-American Tropical Tuna Commission Bulletin* 13: 418–96.

Roy, N. E. Tsoa, and W. Schrank. 1991. "What Do Statistical Demand Curves Show? A Monte Carlo Study of the Effects of Single Equation Estimation of Groundfish Demand Functions." In *Econometric Modelling of the World Trade in Groundfish,* edited by Schrank and Roy. Dordrecht: Kluwer Academic Publishers.

Varian, H. R. 1992. *Microeconomic Analysis.* 3rd ed. New York: Norton and Company.

Wilen, J. E. 1976. "Common Property Resources and the Dynamics of Overexploitation." UBC – PNRE paper no. 3.

World Bank and FAO (Food and Agriculture Organization). 2009. *The Sunken Billions: The Economic Justification for Fisheries Reform.* Washington, DC: World Bank.

The Bio-Economic Model

The bio-economic model

This study's basic bio-economic model follows the modeling standards set in fisheries economics (Anderson 1977; Anderson and Seijo 2010; Clark 1980) with a small addition. In a fairly simple form, it consists of four equations, as follows:

(B.1) $\dot{x} = G(x) - y$ (Biomass growth function)

(B.2) $y = Y(e,x)$ (Harvesting function)

(B.3) $\pi = p \cdot y - C(e)$ (Benefits function)

(B.4) $p = P(x)$ (Landings price function)

The variable x represents the level of biomass and y harvest. The function $G(x)$ represents the natural growth of the biomass before harvesting. Although not explicitly stated, all the variables in this model depend on time. Equation (B.1) describes net biomass growth. In continuous time, $\dot{x} \equiv \partial x/\partial t$, where t denotes time. In discrete time, $\dot{x}$ should be interpreted as $\dot{x} \equiv x(t+1) - x(t)$. Equation (B.2) explains the harvest as a function of fishing effort, e, and biomass, x. Equation (B.3) defines net benefits as the difference between revenues denoted by $p \cdot Y(e,x)$, where p denotes the average net price of landed catch, and costs represented by the cost function $C(e)$. Equation (B.4) is the addition to the standard model. It defines a price function for the landed catch. This equation is supposed to reflect the observation that, as global fish stocks increase, landings will increasingly consist of more valuable species and larger individual fish that typically fetch higher prices (Herrmann 1996; Homans and Wilen 2005).

Of the five variables in this model, that is, x, y, π, p, and e, the first four may be seen as endogenous, that is, they are determined within the fishery. The fifth, fishing effort, or e, may be seen as an exogenous control variable, that is, the variable whose values may be selected to maximize benefits from the fishery. It is easy to check that for any given initial biomass, $x(0)$, say, any selected time path for fishing effort $\{e\}$, say, will generate corresponding time paths for the model's endogenous variables.

The model defined by equations (B.1) to (B.4) may be written more simply and conveniently by recognizing that, provided the harvest function is differentiable and harvest is monotonically increasing in fishing effort, fishing effort can be expressed as the function: $e = E(y, x)$. In that case the entire model can be restated in the following two equations:

(B.1) $\dot{x} = G(x) - y$ (Biomass growth function)

(B.5) $\pi = P(x) \cdot y - \tilde{C}(y, x)$ (Benefits function)

where the new cost function, $\tilde{C}(y, x)$, depends on harvest and biomass. In this form of the basic model, the endogenous variables are biomass, x, and net benefits, π. The exogenous control variable is now harvest, y.

At equilibrium, that is, a sustainable state, the biomass does not change, $\dot{x} = 0$. In that case equations (B.1) and (B.2) define a subsystem in two endogenous variables, x and y, and one exogenous variable, e. From this system it is possible to derive the sustainable harvest and biomass functions as functions of fishing effort only. Since these functions are extremely useful in equilibrium analysis and for illustration, they are worth expressing explicitly as:

         $y = \varphi(e)$ (Sustainable yield function)

         $x = \phi(e)$ (Sustainable biomass function)

It is also worth noting that, at equilibrium, harvest equals biomass growth, and the entire model (B.1 to B.4) may be expressed as

(B.6) $Y(e, x) = G(x)$,

(B.7) $\pi = P(x) \cdot Y(e, x) - C(e)$

where the endogenous variables are now x and π.

Making use of equation (B.5) the same model can be expressed even more succinctly as

(B.8) $\pi = P(x) \cdot G(x) - \tilde{C}(G(x), x)$

These simplified forms of the basic model are extensively used in this study's numerical calculations.

The specific model

The basic model comprises four elementary functions, namely: the natural growth function, $G(x)$, the harvesting function $Y(e, x)$, the cost function, $C(e)$ and the price function, $P(x)$. The specific model is defined by the form of these functions.

The biomass growth function adopted is the Pella-Tomlinson one (Pella and Tomlinson 1969). This function may be written as

(B.9) $G(x) = \alpha \cdot x - \beta \cdot x^{\gamma}$,

where x represents biomass as before and α, β, and γ are parameters. The parameter γ may be referred to as the Pella-Tomlinson exponent. Clearly, for stability, the Pella-Tomlinson exponent must be greater than unity. The Pella-Tomlinson exponent defines the skewness of the biomass growth function. For values of γ that are less than 2 (and greater than unity), the Pella-Tomlinson biomass growth function is skewed to the left. For $\gamma > 2$, the biomass growth function is skewed to the right. For $\gamma = 2$, the biomass growth function is symmetric. There are biological reasons to expect $\gamma \leq 2$ (Branch and others 2013; Pella and Tomlinson 1969; Thorson, Hively, and Hilborn 2012). Figure B.1 illustrates the Pella-Tomlinson function for $\gamma = 1.2$.

Another useful property of the Pella-Tomlinson exponent is that it determines the ratio of the biomass corresponding to the maximum sustainable yield, XMSY,

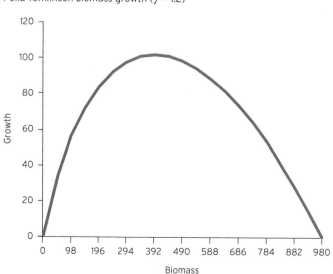

FIGURE B.1
Pella-Tomlinson biomass growth ($\gamma = 1.2$)

say, and the biomass carrying capacity, Xmax. More precisely, for γ ranging from unity to infinity, this ratio ranges between XMSY/Xmax $\in [0.368,1]$ and is 0.5 for $\gamma = 2$. See a later subsection in this appendix for more on the property of γ and its implications.

A great advantage of the Pella-Tomlinson biomass growth function is that it is more flexible than two parameter biomass growth functions such as the Lotka-Volterra logistic function (Volterra 1926) and the Fox function (Fox 1970). In fact, as it is easy to show, the Pella-Tomlinson function incorporates both of these functions as special cases. Thus, if $\gamma = 2$ the Pella-Tomlinson function becomes the Lotka-Volterra logistic function, and as γ gets closer to unity, the Pella-Tomlinson function converges to the Fox biomass growth function.

Some key attributes of the Pella-Tomlinson biomass function expressed as functions of its parameters are listed in table B.1:

TABLE B.1
Attributes of the Pella-Tomlinson biomass growth function

Attribute	Symbol	Expression
Intrinsic growth rate	R	α
Biomass carrying capacity	Xmax	$\left(\dfrac{\alpha}{\beta}\right)^{\frac{1}{\gamma-1}}$
Maximum sustainable yield	MSY	$\alpha \cdot \left(\dfrac{\gamma-1}{\gamma}\right) \cdot \left(\dfrac{\alpha}{\gamma \cdot \beta}\right)^{\frac{1}{\gamma-1}}$
Maximum sustainable yield biomass	XMSY	$\left(\dfrac{\alpha}{\gamma \cdot \beta}\right)^{\frac{1}{\gamma-1}}$
Maximum sustainable yield biomass/ carrying capacity	XMSY/Xmax	$\gamma^{\frac{1}{1-\gamma}}$

For the harvesting function the generalized Schaefer (1954) form is adopted:

(B.10) $Y(e,x) = q \cdot e \cdot x^b$,

where, as before e refers to fishing effort, x biomass, and q and b are positive parameters. This function is monotonically increasing in both fishing effort and biomass. The coefficient q is often referred to as the catchability coefficient. The coefficient b indicates the degree of schooling behavior by the fish. For this reason it is often referred to as the schooling parameter (Bjorndal and Munro 2012). Normally $b \in [0,1]$. If the fish resource is equally distributed over the fishing grounds so that fish search would make no sense, b would be close to unity. If the resource is patchy, that is, the fish form distinct schools of relatively high density, searching for high concentrations makes sense, and b would be smaller than unity, and, depending on the extent of the schooling behavior of the fish, even close to zero.

A schooling parameter less than unity, which is really to be expected for most fish stocks, has major implications for the fishery and the fisheries models. Among other things it leads to bifurcations (sort of discontinuities) in the sustainable yield and biomass curves (Clark 1990). In sustainable fisheries models, these bifurcations correspond to points of fishery collapse which is a major concern for fisheries management.

For the cost function, the following linear form is chosen,

(B.11) $C(e) = c \cdot e + fk,$

where the positive parameter c represents marginal variable costs and fk fixed costs which must be nonnegative.

Finally, the landings price function is defined as

(B.12) $P(x) = a \cdot x^d,$

where a and d are positive parameters. Importantly d is the landings price elasticity with respect to biomass, that is, d measures the percentage increase in the landed catch price as biomass increases by 1 percent.

Assuming biomass equilibrium, that is, a sustainable fishery, it is possible to deduce from the equilibrium or sustainable yield curves as a function of fishing effort for the two biomass growth functions. Figure B.2 provides examples of the corresponding sustainable revenue and cost curves for the model described above. This mode of depicting the fishery is usually referred to as the sustainable fisheries model (see for example, Hannesson [1993]).

FIGURE B.2
An example of the sustainable fisheries model (b = 0.7, γ = 1.2)

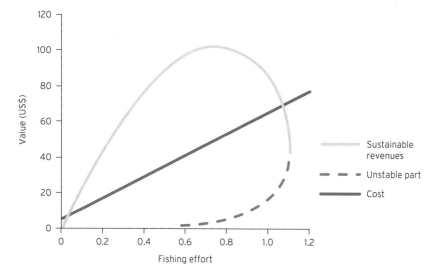

Figure B.2 is drawn for specific values of the model parameters. In particular, the schooling parameter selected is substantially less than unity, more precisely 0.7, and the Pella-Tomlinson exponent, γ, is 1.2 so the biomass growth function is skewed to the left. For any schooling parameter less than unity the sustainable revenue function will have shape similar to the one depicted in figure B.2. Interestingly, if anything, the sustainable revenue function is skewed to the right. More strikingly, it exhibits a bifurcation (or discontinuity) at approximately fishing effort level 1.1 (Clark 1990). Beyond the bifurcation point, there is a backward-bending segment of sustainable revenues that is unstable (dashed curve). This segment has little practical relevance but is theoretically interesting.

From figure B.2 it is obvious that maintaining fishing effort beyond the bifurcation level will eventually lead to the fishery's collapse. Thus the fishery can be sustainable at a significant output level if fishing effort is at less or equal to the bifurcation point, but a small increase in sustained fishing effort beyond this point will eventually lead to the fishery's collapse, again, with major implications for fisheries management.

Sustainable revenue (or yield) functions are traditionally drawn as continuous curves without any bifurcation in the literature on bio-economic models (Anderson 1977; Bjorndal and Munro 2012; Hannesson 1993). Since the example in figure B.2 comes out of a standard fisheries model defined by equations (B.1) to (B.4), this suggests that the traditional way of drawing these curves is overly simplistic and possibly seriously misleading.

As figure B.2 is drawn, fishing costs (curve labeled "cost") intersects the sustainable revenue curve at fishing effort level of just above unity. This intersection is the point toward where an unmanaged fishery would converge and is, therefore, referred to as bionomic equilibrium in the fisheries economics literature (Anderson 1977; Clark 1990). As illustrated, this bionomic equilibrium occurs at the stable point of the sustainable revenue curve. However, as is apparent from the figure, the costs do not have to be reduced (or price increased) by much for the intersection to be on the unstable part of the sustainable revenue curve, in which case the unmanaged fishery would be doomed in the long run. The fact that the bionomic equilibrium is not far away from the unstable part indicates that the risk of a stock collapse is relatively high at the bionomic equilibrium. This is a common feature of many commercial fisheries.

Equilibrium benefits from the fishery are maximized at a fishing effort level where the distance between equilibrium revenues and costs is greatest. As can be seen in figure B.2, this occurs at fishing effort level far less than the one corresponding to bionomic equilibrium.

The reason for the bifurcation of the sustainable revenue curve in this case is that the schooling parameter is less than unity (actually 0.7). The schooling parameter, as already discussed, is far less than unity in many fisheries and would therefore typically be so for an aggregation of several fisheries. Thus, the kind of bifurcation illustrated in figure B.2 would be the normal state of affairs in fisheries and certainly across a collection of many fisheries. It is, however, of great interest, if only to understand the role of the schooling parameter on the sustainable fishery, to wonder about the sustainable revenue function for the schooling parameter, b, equal to unity.

Figure B.3 replicates figure B.2 with the addition that a sustainable revenue function for $b = 1$ is also drawn. This is the only difference between the two revenue functions.

As illustrated in figure B.3, with the schooling parameter, b, equal to unity; there is no bifurcation in the sustainable revenue curve. Therefore, there is no discontinuity or point of stock collapse in this function. Moreover, and related to this, when the schooling parameter is unity, the sustainable revenues are much more resilient at high effort levels than when the schooling parameter is less than unity. Thus, assuming a schooling parameter of unity, a common practice in fisheries modeling, may lead to overly optimistic perceptions of fisheries' resilience to exploitation.

The Pella-Tomlinson exponent (γ)

The Pella-Tomlinson exponent determines the skewness of the Pella-Tomlinson biomass growth function. This exponent, which may be conveniently referred to as γ, cannot be less than unity. For $2 > \gamma \geq 1$, the Pella-Tomlinson biomass growth curve is skewed to the left. This situation means that maximum biomass growth

FIGURE B.3
The sustainable fisheries model: An example with two levels of schooling parameter

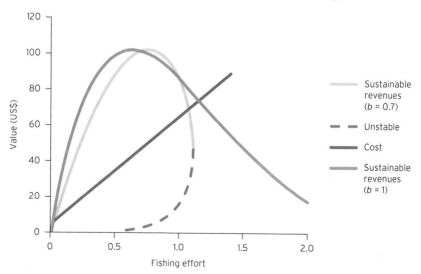

FIGURE B.4
Examples of the Pella-Tomlinson biomass growth function

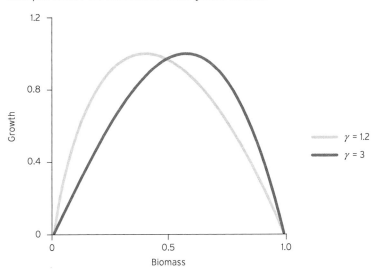

Note: The graph shows the same maximum sustainable yield and carrying capacity, where γ = 1.2 and γ = 3.

and therefore maximum sustainable yield occur at biomass less than half of the carrying capacity of the biomass (Thorson and others 2012). For $\gamma > 2$, the Pella-Tomlinson biomass growth curve is skewed to the right. Figure B.4 highlights examples of the Pella-Tomlinson biomass growth function for $\gamma < 2$ and $\gamma > 2$.

Clearly if the biomass growth function is skewed to the left, biomass growth will be comparatively high at relatively low biomass levels. This result implies that the fishery will be more resilient to a high fishing effort than would otherwise be the case.

To illustrate this, two sustainable yield functions corresponding to the biomass growth functions in figure B.4 are drawn in figure B.5.[1] Both functions exhibit the same maximum sustainable yield (MSY), in accordance with the two biomass growth functions in figure B.4. However, this MSY is attained at a significantly higher fishing effort for the biomass growth function that is skewed to the left (γ = 1.2) than the one skewed to the right (γ = 3). Moreover, the resilience of the former to a high fishing effort is much greater than that of the latter. It is also worth noting that the sustainable yield function corresponding to the biomass growth function with γ = 3 exhibits a discontinuity at approximately a 2.6 fishing effort level.

As suggested by figure B.5, the value of the Pella-Tomlinson exponent, γ, has major implications for fisheries policy and management. The higher the value of γ, the less resilient the fishery will be to fishing effort, and vice versa. Thus, the risk of a stock collapse increases with the value of γ, all else being equal.

FIGURE B.5
Sustainable yield functions

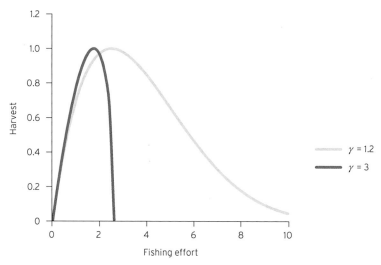

Note: γ = 1.2 and γ = 3, and harvesting $y = e \cdot x$.

There is a great deal of indirect evidence that fisheries may commonly be described by a Pella-Tomlinson exponent, γ, that is substantially less than two. Many fisheries, especially those that have been intensely exploited for a long time, have shown a remarkable resilience to a persistently high fishing effort. Although, undoubtedly, many other factors are at play in this process, one reason could be a biomass growth function skewed to the left, which can be represented by the Pella-Tomlinson exponent being less than two.

A stylized example of the impact of a larger stock on the average price of fish

Global fish stocks are severely overexploited. The most highly valued stocks, which are usually at a high trophic level, are much more overexploited than the others. Thus, restoring global fish stocks to their economically optimal level implies that the most valuable stocks will increase more than the less valuable stocks. In fact, for ecosystem reasons, some of the less valuable stocks may actually decline. This reasoning implies that if global fish stocks are restored, the share of more valuable fish in total landings will be increased, meaning that the average price of landed volume will increase.

For obvious reasons (the empirical evidence is simply unavailable) not much is known about the quantitative impact of global fish stocks on the average price of landed catch. The following numerical example attempts to provide some idea of the magnitudes involved.

Divide the global fishery into two stocks. One is at a high trophic level and highly priced. The other is at a low trophic level and fetches a low price. Assume that initially both stocks are overexploited, the valuable stock more so than the less valuable one. Then initiate an effort to rebuild the stocks, which leads to doubling of the aggregate biomass, through a reduction in exploitation rates.

TABLE B.2
A stylized example of the impact of stock increase on average landings price

Stocks	Biomass $x(0)$	$x(1)$	Unit price p	Exploitation rate $f(0)$	$f(1)$	Harvest $y(0)$	$y(1)$	Landed value $v(0)$	$v(1)$
More valuable	0.3	1.0	2.0	0.30	0.20	0.09	0.20	0.2	0.4
Less valuable	0.7	1.0	1.0	0.40	0.20	0.28	0.20	0.3	0.2
Aggregate	1.0	2.0		0.37	0.20	0.37	0.40	0.5	0.6

For this purpose, we assume reasonable fishing mortality rates before and after the stock rebuilding.

The following table summarizes the example:

From this table, it is straightforward to calculate the average price of landed catch before and after the stock restoration, namely 1.2 and 1.5, respectively. The increase is 20.7 percent, and the elasticity of price with respect to biomass is 0.207.

It should be noted that this example does not include the impact of the increased proportion of larger individuals in the aggregate catch which can be substantial.

Note

1. It should be noted that these sustainable yield functions are drawn for a simple harvesting function with a schooling parameter equal to unity. For schooling parameter less than unity, the shape of the sustainable function changes but their essential skewness and relative resilience to fishing effort remains.

References

Anderson, L. 1977. "The Economics of Fisheries Management." Johns Hopkins University Press, Baltimore.

Anderson, L. and J. Seijo. 2010. *Bioeconomics of Fisheries Management.* New York: John Wiley & Sons.

Bjorndal, T., and G. Munro. 2012. *The Economics and Management of World Fisheries.* U.S.: Oxford University Press.

Branch, T., D. Hively, and R. Hilborn. 2013. "Is the Ocean Food Provision Index Biased?" *Nature* 495: E5–6.

Clark, C. W. 1980. "Towards a Predictive Model for the Regulation of Fisheries." *Canadian Journal of Fisheries and Aquatic Science* 37: 1111–129.

Clark, C. 1990. *Mathematical Bioeconomics: The Optimal Management of Renewable Resources.* 2nd ed. New York: John Wiley & Sons.

Fox, W. 1970. "An Exponential Surplus Model for Optimizing Exploited Fish Populations." *Transactions of the American Fisheries Society* 99: 80–88.

Hannesson, R. 1993. "Bioeconomic Analysis of Fisheries." Food and Agriculture Organization of the United Nations, Rome.

Herrmann, M. 1996. "Estimating the Induced Price Increase for Canadian Pacific Halibut with the Introduction of Individual Vessel Quota System." *Canadian Journal of Agricultural Economics* 44: 151–64.

Homans, F. and J. Wilen. 2005. "Markets and Rent Dissipation in Regulated Open Access Fisheries." *Journal of Environmental Economics and Management* 49: 381–404.

Pella, J. J., and P. K. Tomlinson. 1969. "A Generalized Stock Production Model." *Inter-American Tropical Tuna Commission Bulletin* 13: 418–96.

Schaefer, M. B. 1954. "Some Aspects of the Dynamics of Populations Important to the Management of Commercial Marine Species." *Inter-American Tropical Tuna Commission Bulletin* 1: 27–56.

Thorson, J. J. Cope, T. Branch, and O. Jensen. 2012. "Spawning Biomass Reference Points for Exploited Marine Fishes, Incorporating Taxonomic and Body Size Information." *Canadian Journal of Fisheries and Aquatic Science* 69: 1556–68.

Volterra, V. 1926. "Fluctuations in the Abundance of a Species Considered Mathematically." *Nature* 118: 558–56.

APPENDIX C

Estimation of Model Inputs

The "specific" bio-economic model defined in appendix B (equations (B.9) to (B.12)) contains nine unknown coefficients. These coefficients as well as the a priori restrictions on their values are listed in table C.1. To make the model operational, these have to be estimated.

The model also contains six variables: biomass, x; harvest, y; fishing effort, e; global fishery profits, π; and average landings price, p; as well as the change in biomass, $\dot{x}$. With time series data on these variables, in principle, it would be possible to estimate statistically the unknown coefficients listed in table C.1. Unfortunately, reasonably reliable time series data are available for only one of these variables, the global harvest, y. For this reason the conventional statistical estimation procedure was not feasible. Instead, the estimates were derived by combining the best available empirical inputs, that is, estimates of global fisheries parameters and fisheries variables in a specific base year (see details following).

The empirical input data employed are summarized in table C.2, and estimates of empirical model inputs are in table C.3.

Global maximum sustainable yield (MSY)

When determining the global MSY and other biological reference points of the global fishery, it is crucial to be clear about the fish stocks that are being considered. In this instance, the fish stocks covered are marine stocks that have

TABLE C.1
Model coefficients and variables that need to be estimated

Coefficients	Symbol	Characterization	Permissible values
Biological coefficients			
Biomass growth function	α	Intrinsic growth rate	$a > 0$
Biomass growth function	β		$b > 0$
Biomass growth function	γ	Pella-Tomlinson exponent	$g > 1$
Bio-economic coefficients			
Harvesting function	q	Catchability	$q > 0$
Harvesting function	b	Schooling parameter	$0 < b \leq 1$
Economic coefficients			
Cost function	c	Marginal cost parameter	$c > 0$
Cost function	fk	Fixed costs	$fk \geq 0$
Price function	a	Landings price parameter	$a > 0$
Price function	d	Elasticity of price w.r.t. biomass	$d \geq 0$

Note: w.r.t. = with respect to.

TABLE C.2
Global data for estimation of model coefficients and base-year variables

Biological data	Symbol
Maximum sustainable yield	MSY
Carrying capacity	Xmax
Pella-Tomlinson exponent	γ
Schooling parameter	b
Economic data	
Fixed costs	fk
Elasticity of landings price with respect to biomass	d
Base-year (t_0) variables	
Landed quantity	$y(t_0)$
Average landing price	$p(t_0)$
Biomass growth	$\dot{x}(t_0)$
Profits	$\pi(t_0)$

a history of significant exploitation, which excludes very large stocks of fish that have hardly been exploited, such as Antarctic krill and lanternfish.[1]

While there are different estimates of the global marine MSY, *as understood above*, the most prominent estimates range between 80 and 115 million tons (Alverson and Dunlap 1998). The initial *Sunken Billions* publication (World Bank and FAO 2009) adopted 95 million tons as its "conservative" estimate, while more recent estimates range between 83 and 99 million tons (Sumaila and others 2012) and 85 to 110 million tons (Costello and others 2012).

The highest reported marine catch that occurred in 1996 is just above 86 million tons (FAO 2014). The subsequent decline of global fish stocks suggests that this level exceeded the global MSY. Even when accounting for unreported catches and discards of less valuable catches, it is unlikely that marine catches of valuable species have ever exceeded 100 million tons. This result can be taken to suggest that the global MSY is less than 100 million tons.

However, there are reasons to believe that the historical path of marine harvests underestimates the global MSY, and for this reason, the slightly higher global MSY estimate of 102 million tons is used here. The main reason is that the global fishery consists of a large number of fish stocks, and these stocks have been exploited sequentially, with the most easily accessible and valuable stocks fished first. Thus, during the historic process of global marine fishing, exploitation of many species may have exceeded the MSY of individual stocks, but since this shift occurred sequentially, the catch from all exploited species may never have actually exceeded the global MSY.

The carrying capacity (Xmax)

While the global marine fisheries MSY is difficult to determine, even less is known about the global carrying capacity of currently exploited fish stocks. Fortunately, sensitivity studies indicate that its impact on the ultimate estimate of sunken billions is largely limited.

For different species of fish, the carrying capacity of the stock is between 5 and 15 times the MSY. In fact, for the Pella-Tomlinson biomass growth function with reasonable values of the parameters, this multiple would typically be between 9 and 12. Accordingly, the global carrying capacity is set at 980 million tons, which is about 9.6 times the assumed MSY.

The Pella-Tomlinson exponent (γ)

This parameter determines the skewness of the Pella-Tomlinson biomass growth function (see appendix B). Thorson and his associates recently conducted a major empirical study of the value of γ in a number of fisheries around the world (Thorson and others 2012). Based on a database of landings and biomass for 147 fish stocks covering the main types of commercial fish stocks fairly widely distributed around the world, that study found that larger carnivorous fish (for example, gadoids and many demersals) typically exhibited a higher value of γ, while smaller plankton feeding fish (for example, herrings and anchovies) exhibited a lower value of γ. For all stocks taken together, they found that the MSY occurred approximately at 40 percent of the carrying capacity of these stocks. This situation corresponds to a Pella-Tomlinson exponent, γ, approximately equal to 1.188, which was used in this study.

Schooling parameter (*b*)

This parameter appears in the harvesting function and characterizes the schooling behavior of the stocks, which in turn affects fishing efficiency. For fish species with a strong tendency to congregate in relatively dense schools or shoals (such as herrings, anchovies and sardines), harvest levels are often little influenced by the overall abundance of the stock (Bjorndal and Munro 2012; Clark 1990; Hannesson 1993). The opposite is true for species that are relatively uniformly distributed over the fishing grounds (such as cod or sharks). For these species, harvests tend to vary proportionately with the available biomass for any given level of fishing effort.

The so-called schooling parameter attempts to capture these features of the species. The parameter normally has a value between zero and unity. The lower the schooling parameter, the more pronounced the schooling behavior, and the less dependent is the harvest on biomass. For many commercial species (for instance, many bottom dwelling, or demersal species and shellfish), it is typically close to unity (Arnason 1984, 1990). For pelagic species (such as tuna, herring, and sardine), it is often much lower (Bjorndal 1987; Bjorndal and Munro 2012).[2]

In the harvesting function of the bio-economic model for the global fishery, the schooling parameter should reflect the average schooling behavior of different fish species. The schooling parameter adopted in this study is the weighted average of the main marine species groups using their shares in the total volume of catch as weights.[3]

As shown in figure C.1, the aggregate schooling parameter has been remarkably constant at around 0.70 since 1970. Since 2000, the stability has increased and a slight upward trend can be detected.

FIGURE C.1
Evolution of estimated aggregate schooling parameter, 1970–2012

Source: Global data aggregated by the authors.
Note: MSY (maximum sustainable yield) values are the historical maximum catches by species group as reported by FAO Fishstat Plus. The schooling parameters are assumed, based on information on schooling parameters for several indicative species.

The long-term average of the aggregate schooling parameter is just over 0.72, while the 2012 estimated value is just under 0.72. Therefore, in this study, a global schooling parameter of 0.72 is adopted.

Elasticity of price with respect to biomass (*d*)

The average landings price depends on the global biomass of commercially exploited marine fish stocks in accordance with a parameter referred to as the elasticity of price with respect to biomass. The reasons for this were discussed in chapter 2: fishing activities usually target the most valuable fish stocks first.[4] These are high-value species and tend to be at a high trophic level, that is, high on the food web of marine ecosystems. As fishing effort increases, the most valuable stocks become relatively scarce and the fishing activity moves down to the next valuable fish stocks. This phenomenon is known as "fishing down the marine food web" (Pauly and others 1998).

This study uses a value of 0.22 for the price elasticity parameter, which means that, if the global biomass doubles (a 100 percent increase), then the average price of landings increases by 22 percent, with 0.22 a conservative estimate of the elasticity of global price in relation to fish stocks level.

Biomass growth (*x̂*) in the base year

According to the Food and Agricultural Organization of the UN (FAO) statistics (FishStat Plus; FAO 2014), and after a significant upward trend for decades, global marine catches have been relatively constant since the early 1990s, fluctu-ating between about 78 and 86 million tons. This is consistent with the aggregate global biomass stabilizing over the same period. During the latter part of this period, however, and especially since 2004, global marine catches have exhibited a slow declining trend. During the 10 years between 1993 and 2002, the average marine catch was about 83 million tons with no apparent upward or downward trend, while the average between 2003 and 2012 was 80.5 million tons with a weak downward trend. It is likely, however, that this decline in global catches was caused more by attempts by several major fishing nations to restore fish stocks (through catch reduction efforts), rather than actual declining fish stocks.

Although it appears that stocks have continued declining in some parts of the world (especially Southeast Asia and Africa), there is some evidence that stocks in other regions may have actually improved, sometimes substantially. Thus, although the evolution of fish stocks is far from uniform across the world and declines undoubtedly prevail in some regions, it appears that, in our 2012 base year, net global fish stocks may actually have increased, led by increases in the North Atlantic and Pacific. For the purpose of this study, the 2012 biomass increase input is estimated at around 8 million tons. While it may at first appear large, this volume of fish is actually quite small relative to the biomass level that supports maximum economic yield (MEY), not to mention the global carrying capacity of all commercially exploited stocks. Nevertheless, this degree of biomass growth represents a major change from the original *Sunken Billions* study, where the global biomass was estimated to have declined by 2 million tons in the year 2004.

Volume of landings (y) in the base year

FAO's FishStat Plus is the only comprehensive global database on fish landings that is currently available to public. In accordance with the official FAO statistics, the global marine landings of 79.6 million tons are used as the 2012 base-year value of y in this study.

The FAO FishStat Plus database has its own limitations (FAO 1999), since it relies on official catch statistics, as provided by individual member countries. In the past, country reports have been doubted for several reasons, including deliberate under- and over-reporting by some authorities, widespread underestimation of production by small-scale fishers, discards at sea, and underreporting of catches by fishers at landing sites (MRAG and UBC Fisheries Centre 2008; World Bank and FAO 2009; Zeller and Pauly 2007). In any case, and in the absence of a robust basis for adjusting the FAO global marine landings statistics as they stand, the FAO FishStat Plus estimate is adopted in this study.

Price of landed catch (p) in the base year

From the point of capture, catches are subject to several stages of handling and processing, first aboard the fishing vessel and subsequently on land where a chain of handling, processing, and distribution eventually brings fish products to the end user. Along this chain, often referred to as the supply chain (Lem 2015), value is generally added and different prices apply at various points. This study, however, is concerned with the strict economics of fishing, and not that of processing or distribution of landed fish. The focus is thus on the price of fish in as unprocessed a form as possible, which is usually understood to be at the point of landing.

There are limited available data on the landed catch price worldwide. While the volume of landings is often poorly monitored and registered, the landed catch price is even less well known, as it is even more difficult to monitor than the volume of landings. This is due, in part, to the fact that the landings price is often not well defined. This lack of price definition can be explained by (i) the reported selling price of landed catch varies according to the different stages of onboard processing (whole, gutted, headed and gutted, filleted and frozen, and so on); (ii) the reported landings price may incorporate varying portions of costs of landing, handling, packaging, and even processing; (iii) the reported landings price may include fees and taxes; (iv) the landed catch quality varies tremendously, particularly depending on the type of boat on which the fish was caught, and consequently the price as well; and (v) there is often vertical integration of fishers and the first receivers of landed catch where reported prices at landing are not set explicitly and/or are artificially distorted. For all these reasons, reasonably comprehensive and reliable datasets for the landed catch price are not globally available.

The average landed catch price used in this study is based on several sources. The single most important source is the FAO assessment of the global value and volume of landed catch by species groups (FAO FishStat Plus). Figure C.2 shows the evolution of the implied average landings price. The data show a consistent increase in the nominal price of catch (except in year 2009), while the real price was stable in the latter half of the 2000s, followed by a steep increase in the early 2010s.

Information about the landed catch value and price in individual fisheries around the world is also used to verify and in some cases adjust the unit value implied by FAO statistics. An important source of price information at the individual fisheries level is the fisheries database compiled by C. Costello and his associates at the University of California Santa Barbara, which covers over 4,000 individual fisheries (Costello and others 2012, 2015). Based on the analysis, the global average price of landed marine catch entered in the model for 2012 is $1,290 per ton.

FIGURE C.2
Implied average landed catch price

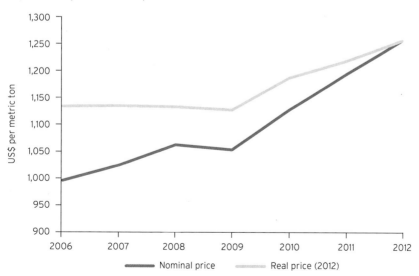

Source: FAO FishStat Plus database.
Note: Real price in 2012 US$ calculated using the U.S. Consumer Price Index.

Fisheries net benefits (π) in the base year

Net benefits of the global fishery in the base year are an important input into the global bio-economic model used in this study. The term "net benefits" refers to the benefits generated by the fishing activity, and as such does not necessarily coincide with accounting *profits*. (See appendix A for further discussion of the concept of net benefits.)

Fishing cost data are highly variable and difficult to come by. At one extreme are the European Union (EU) and Nordic nations, which systematically collect and publish information about the operating results of their fishing fleets. Among these countries are the EU nations (see European Commission Joint Research Centre), Norway (see Fiskeridirektoratet i Norge), and Iceland (see Statistics Iceland). In some countries, especially in the EU, fisheries profitability was found to be negative in 2012, although overall fisheries profits in the EU were estimated to be slightly positive (Paulrud, Carvalho, and Borrello 2014).

In many developing countries, data are lacking or inadequate. In an attempt to fill this gap, the authors conducted a survey of fishing costs and profitability by means of personal communications with national experts around the world. Broadly speaking, fisheries profitability seems poor, although rarely highly negative, and with great variability between fisheries and countries.

Fishing costs

According to the data collected and other available information, the structure of fishing costs across fisheries and nations is broadly similar.[5] The most important cost categories are: (i) labor costs, (ii) fuel costs, (iii) other operating costs including fishing gear and maintenance, and (iv) capital costs, which consist essentially of interest payments and capital depreciation. Figure C.3 shows the

FIGURE C.3
Cost structure in the fishing industry, 2012

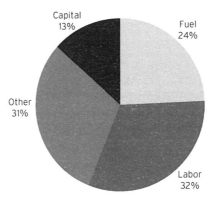

Sources: FAO 1993; Joint Research Centre 2014; Iceland Statistics 2014; World Bank and FAO 2009.
Note: Share in total costs.

share of these cost categories in total costs for 2012. As the pie chart shows, labor remuneration is the single most important cost item, accounting for almost one-third of total costs, followed by fuel and capital costs. Other costs, consisting of a collection of various smaller cost items including fishing gear and maintenance, are almost one-third of total costs.

Unlike most other cost items, labor remuneration is determined as a percentage of the value of landings in most fisheries in the world, although sometimes certain operating costs are subtracted. Thus, fishing workers generally assume a share of the risks of the fishing operation, even when they do not own a share of the fishing capital, such as the vessel or fishing gear. From the vessel owner's perspective, labor costs appear primarily as a reduction in value of the landed catch. This virtually universal arrangement in commercial fishing is likely to have implications on the behavior of fishing fleets as rational economic actors.

Adjusted benefits

Net benefits in global fisheries increased slightly between 2004 and 2012, not including changes in taxes and subsidies. These higher net benefits can be explained by the advent of more efficient fisheries management systems and additional improvements that have helped tackle the severe state of overcapacity. However, there is some evidence that modestly reduced subsidies and higher taxes during this period offset total fishery sector financial gains—although the magnitude is insufficient to alter basic trends of overfishing.

Model Specification

Combining the above empirical data with the structure of the bio-economic model leads to the estimation formulae listed in table C.4. These formulae are

TABLE C.3
Estimates of empirical model inputs

Biological data	Symbol	Values	Units
Maximum sustainable yield	MSY	102	Million tons
Carrying capacity	Xmax	980	Million tons
Pella-Tomlinson exponent	γ	1.188	None
Schooling parameter	B	0.71	None
Economic data			
Fixed costs	fk	0	US$ billion
Elasticity of landings price with respect to biomass	d	0.22	None
Base-year (2012) variables			
Landed quantity in 2012	$y(t_0)$	79.7	Million tons
Average landing price in 2012	$p(t_0)$	1.29	US$/kg
Biomass growth in 2012	$\dot{x}(t_0)$	8	Million tons
Net benefits in 2012	$\pi(t_0)$	3	US$ billion

TABLE C.4

Formulae to calculate model parameters and base-year variables

Unknowns	Estimation formulae	Comments
Biological coefficients		
α	$\hat{\alpha} = \left(\dfrac{MSY}{Xmax}\right) \cdot \left(\dfrac{\gamma^{\frac{\gamma}{\gamma-1}}}{\gamma-1}\right)$	
β	$\hat{\beta} = \left(\dfrac{MSY}{Xmax^{\gamma}}\right) \cdot \left(\dfrac{\gamma^{\frac{\gamma}{\gamma-1}}}{\gamma-1}\right)$	
q	$\hat{q} = \dfrac{y(t_0)}{e(t_0) \cdot x(t_0)^b}$	
Economic coefficients		
c	$\hat{c} = \dfrac{p(t_0) \cdot y(t_0) - \pi(t_0) - fk}{e(t_0)}$	
A	$\hat{a} = \dfrac{p(t_0)}{\hat{x}(t_0)^d}$	
Base-year variables		
$x(t_0)$	$\hat{G}(\hat{x}(t_0)) = y(t_0) + \dot{x}(t_0)$	Nonlinear numerical search
$e(t_0)$	1	Normalization

derived from the properties of the Pella-Tomlinson biomass growth function and the basic structure of the bio-economic model.

The only significant complication in applying the estimation formulae in table C.5 concerns the estimation of base-year biomass. As the graph of the Pella-Tomlinson biomass growth function, figure B.1, makes clear, the biomass growth function is non-monotonic. Therefore, the estimation formula for $x(t_0)$ will usually produce two nonnegative estimates of the base-year biomass. One of them will be larger than the biomass corresponding to the maximum sustainable yield, XMSY. The other will be larger. Extraneous information about the state of the global fishery will have to be used to determine which of these two possible estimates is appropriate.

The resulting estimates are summarized in table C.5.

On the basis of these empirical specifications, the global fisheries model can be illustrated graphically as in figure C.4. In this diagram, the orange square indicates the approximate location of the global fishery in the 2012 base year.

TABLE C.5
Model coefficients and base-year variables

	Characterization	Values	How obtained
Biological coefficients			
α	Intrinsic growth rate	1.644	Calculated
β		0.45	Calculated
γ	Pella-Tomlinson exponent	1.188	Estimated
Bio-economic coefficients			
Q	Catchability	1.76	Calculated
B	Schooling parameter	0.71	Estimated
Economic coefficients			
C	Marginal cost parameter	97.4	Calculated
fk	Fixed costs	0	Inferred
a	Landings price parameter	0.39	Calculated
d	Elasticity of price with respect to biomass	0.22	Estimated
Base-year (2012) variables			
$y(2012)$	Landed quantity	79.7	Estimated
$p(2012)$	Average landing price	1.26	Estimated
$\dot{x}(2012)$	Biomass growth	8	Estimated
$\pi(2012)$	Net benefits	3	Estimated
$e(2012)$	Fishing effort	1	Normalized
$x(2012)$	Biomass	214.9	Calculated

FIGURE C.4
The sustainable fishery, 2012

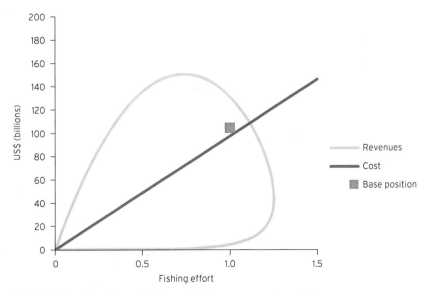

Note: The backward bending part of the sustainable revenue curve is unstable.

Notes

1. The biomass of Antarctic krill may be 300–500 million tons (Atkinson et al. 2009) and that of lanternfish could easily be 600 million tons (Hulley 1996). The combined biomass of only these two species exceeds that of the whole of the commercially exploited stocks covered in this study. So far, these two, and many similar stocks, have only been very modestly exploited, and their inclusion in the modeling alone would raise the global MSY possibly by tens of millions of tons.

2. A schooling parameter of less than unity leads to a discontinuity in the sustainable yield and revenue functions at some level of fishing effort (appendix B). These discontinuities are of concern because they correspond to a fisheries collapse if fishing effort is maintained above that critical level for a sufficiently long time.

   Another potentially significant factor affecting fishing efficiency in a similar way is the fish stock size for a given species. As stocks grow, catch volume per unit fishing time will typically increase.

3. Demersal: 1.0; pelagic: 0.6; cephalopods: 0.7; mollusks: 1.0.

4. Undesirable effects of selective fishing on ecosystems are discussed in Garcia and others (2012).

5. However, great variations exist within a fishery, for example, between artisanal and industrial fishing fleets.

References

Alverson, D., and K. Dunlop. 1998. "Status of the World Marine Fish Stocks." Fisheries Research Institute, School of Fisheries, University of Washington, Seattle.

Arnason, R. 1984. "Efficient Harvesting of Fish Stocks: The Case of the Icelandic Demersal Fisheries." PhD dissertation, University of British Columbia.

———. 1990. "A Numerical Model of the Icelandic Demersal Fisheries." In *Operations Research in Fisheries and Fisheries Management,* edited by G. Rodriguez. Kluwer.

Atkinson, A., V. Siegel, E. Pakhomov, M. Jessopp, and V. Loeb. 2009. "A Re-Appraisal of the Total Biomass and Annual Production of Antarctic Krill." *Deep-Sea Research* I 56: 727–40.

Bjorndal, T. 1987. "Production Economics and Optimal Stock Size in a North Atlantic Fishery." *Scandinavian Journal of Economics* 89: 145–64.

Bjorndal, T., and G. Munro. 2012. *The Economics and Management of World Fisheries.* U.S.: Oxford University Press.

Clark, C. 1990. *Mathematical Bioeconomics: The Optimal Management of Renewable Resources.* 2nd ed. New York: John Wiley & Sons.

Costello, C., D. Ovando, T. Clavelle, C. Strauss, R. Hilborn, M. Melnychuk, T. Branch, S. Gaines, C. Szuwalski, R. Cabral, D. Rader, and A. Leland. 2015. "Have Your Fish and Eat Them Too." Unpublished manuscript.

Costello, C., D. Ovando, R. Hilborn, S. Gaines, O. Deschenes, and S. Lester. 2012. "Status and Solutions for the World's Unassessed Fisheries." *Science* 338: 517–20.

FAO (Food and Agriculture Organization of the UN). 1993. "Marine Fisheries and the Law of the Sea: A Decade of Change." In *The State of Food and Agriculture 1992* (special chapter, revised). FAO Circular No. 853. FAO, Rome.

———. 2014. "The State of World Fisheries and Aquaculture 2014 (SOFIA)." FAO, Rome.

FAO Fishstat Plus (database). http://www.fao.org/fishery/statistics/software/fishstat/en. FAO, Rome.

Garcia, S. M., J. Kolding, J. Rice, M.-J. Rochet, S. Zhou, T. Arimoto, J. E. Beyer, L. Borges, A. Bundy, D. Dunn, E. A. Fulton, M. Hall, M. Heino, R. Law, M. Makino, A. D. Rijnsdorp, F. Simard, and A. D. M. Smith. 2012. "Reconsidering the Consequences of Selective Fisheries." *Science* 335 (March): 1045–77.

Hannesson, R. 1993. "Bioeconomic Analysis of Fisheries." Food and Agriculture Organization of the United Nations, Rome.

Hulley, P. 1996. "Lantern Fishes." In *Encyclopedia of Fishes*, edited by J. Paxton and W. Eschmeyer. London: Academic Press. Joint Research Centre: https://ec.europa.eu/jrc.

Iceland Statistics. http://www.statice.is/statistics/business-sectors/fisheries/.

MRAG and UBC Fisheries Centre. 2008. "The Global Extent of Illegal Fishing: Final Report." Marine Resources Assessment Group and University of British Columbia Fisheries Centre, Vancouver.

Paulrud, A., N. Carvalho, and A. Borrello, eds. 2014. "The 2014 Annual Report on the EU Fishing Fleet." *JRC Scientific and Policy Reports*. Ispra, Italy.

Pauly, D., V. Christensen, J. Dalsgaard, R. Froese, and F. Torres. 1998. "Fishing Down Marine Food Webs." *Science* 279 (5352): 860–63.

Sumaila, R., W. Cheung, A. Dyck, K. Gueye, L. Huang, V. Lam, D. Pauly, T. Srinivasan, W. Swartz, and R. Watson. 2012. "Benefits of Rebuilding Global Marine Fisheries Outweigh Costs." PLoS ONE 7: e40542.

Thorson, J. J. Cope, T. Branch, and O. Jensen. 2012. "Spawning Biomass Reference Points for Exploited Marine Fishes, Incorporating Taxonomic and Body Size Information." *Canadian Journal of Fisheries and Aquatic Science* 69: 1556–68.

World Bank and FAO (Food and Agriculture Organization). 2009. *The Sunken Billions: The Economic Justification for Fisheries Reform*. Washington, DC: World Bank.

Zeller, D., and D. Pauly, eds. 2007. "Reconstruction of Marine Fisheries Catches for Key Countries and Regions (1950–2005)." Fisheries Centre Research Reports 15(2): 163, The Fisheries Centre, University of British Columbia, Vancouver.